"十三五"普通高等教育本科部委级规划教材

纺织服装基础英语
Textile and Fashion Basic English

李思龙　主编
沈梅英　施慧敏　副主编

U0189817

国家一级出版社　中国纺织出版社　全国百佳图书出版单位

内容提要

本教材是"十三五"普通高等教育本科部委级规划教材。全书共八个单元，内容涵盖纺织服装历史、纺织服装专业基础知识、纺织服装国际贸易等方面。各单元主题明确，文章易懂，标注详尽，习题全面，强调基础，重视实用。教材的配套材料包括听力音频及其文字材料和听力练习答案、每篇文章的练习参考答案。

本教材适用于已完成大学英语基础阶段教学的纺织类、服装类、艺术类等专业的二年级或高年级学生使用，是大学英语拓展课程教材、后续课程教材和专门用途英语（ESP）教材。本教材把纺织基础英语和服装基础英语的学习融入英语语言教学中，在听、说、读、写、译等方面努力提高学生的专业基础英语水平。

图书在版编目（CIP）数据

纺织服装基础英语 / 李思龙主编. --北京：中国纺织出版社，2017.10

"十三五"普通高等教育本科部委级规划教材
ISBN 978-7-5180-3929-6

Ⅰ.①纺… Ⅱ.①李… Ⅲ.①纺织工业—英语—高等学校—教材②服装工业—英语—高等学校—教材 Ⅳ.①TS1②TS941

中国版本图书馆CIP数据核字（2017）第204784号

策划编辑：陈静杰　　责任校对：寇晨晨
责任设计：何　建　　责任印制：王艳丽

中国纺织出版社出版发行
地址：北京市朝阳区百子湾东里A407号楼　邮政编码：100124
销售电话：010—67004322　传真：010—87155801
http://www.c-textilep.com
E-mail: faxing@c-textilep.com
中国纺织出版社天猫旗舰店
官方微博http://weibo.com/2119887771
三河市宏盛印务有限公司印刷　各地新华书店经销
2017年10月第1版第1次印刷
开本：787×1092　1/16　印张：13.25
字数：240千字　定价：49.80元

凡购本书，如有缺页、倒页、脱页，由本社图书营销中心调换

编委会名单

主　编　李思龙
副主编　沈梅英　施慧敏
编　委　（以姓氏拼音为序）
　　　　傅　霞　高　歌　黄　慧　李思龙
　　　　阮　瑾　沈梅英　施慧敏　杨　柳
　　　　张劲松　张　阳　朱　佳　朱　赟

前言

本书是大学英语拓展课程教材和后续课程教材，也是大学专门用途英语（ESP）教材，适合已完成大学英语基础阶段学习的材纺类、服装类、艺术类等专业的大学二年级或高年级学生以及对这些专业感兴趣的其他专业学生使用。

本书将纺织基础英语和服装基础英语融为一体，语言教学和专业基础英语教学相结合，强调语言知识在专业基础英语中的应用，使学生在学习语言的同时了解纺织和服装专业基础知识，在听、说、读、写、译等方面得到全面的提高。

本书共分八个单元，内容涵盖纺织和服装历史、纺织纤维、纺纱技术、织造技术、针织技术、服装设计、服装营销、纺织和服装国际贸易。本书附有光盘，内容包括听力音频以及音频文字材料、听力练习答案和各单元课后练习参考答案。

本书的主要特点：

（1）主题明确，内容丰富：每个单元紧扣同一主题展开，内容包括听力材料和三篇文章，听力材料和文章后都配有相应的思考题和练习题，注重提高学生听、说、读、写、译等语言技能，有利于培养学生的语言能力和文化素养。

（2）文章易懂，图文并茂：材料语言地道，贴近生活，大部分文章都配有图表，这些图表使呆板的文字变得生动，使复杂的工艺变得简单，加深学生对语篇的理解，提高学生对专业基础英语学习的兴趣。

（3）标注详尽，习题全面：听力材料和文章后面都附有详尽的词汇以及音标、词性、中文注释等，有利于读者快速查阅生词、理解生词或短语在文章中的含义。课文后有重点句和难句的英汉对照，还有大量的课外阅读题、词汇题、翻译题等。

（4）强调基础，重视实用：强调专业基础性和内容实用性。通过学习，学生对纺织、服装专业有初步的认识和了解，提高对纺织、服装专业学习的兴趣，为学习高年级的纺织、服装专业英语打下坚实的基础。

通过本书的讲授和训练，要求学生达到以下目标：

（1）初步了解纺织和服装发展史以及相关的历史人物。

（2）掌握纺织服装英语中的基本词汇。

（3）初步掌握科技文章的阅读技巧和方法。

（4）提高学生学习纺织服装英语的兴趣。

（5）顺利过渡到高年级的纺织、服装专业英语学习。

本书由浙江理工大学、东华大学等高校老师编写，各单元参编人员如下：第一单元傅霞、李思龙；第二单元李思龙、沈梅英；第三单元阮瑾、朱佳；第四单元高歌、张劲松；第五单元杨柳、李思龙；第六单元黄慧、张阳；第七单元朱赟、沈梅英；第八单元施慧敏、李思龙。在此，对编写组全体成员的通力合作深表谢意，同时感谢美国英语语言专家Donald E.Moreale的支持和帮助。此外，本书在编写过程中参考了众多学者和行业专家的经验，在此深表敬意和谢意。由于编者水平有限，本教材的疏漏和错误之处，欢迎大家批评指正。

编　者
2017年3月

Contents

Unit 1   History of Textile and Fashion ········· 001
    PART ONE   Warm-up Activities ········· 001
    PART TWO   Reading Activities ········· 003
        Passage 1   History of Fabric and Textile ········· 003
        Passage 2   History of Silk ········· 013
        Passage 3   History of Jeans ········· 019

Unit 2   Textile Fibers ········· 025
    PART ONE   Warm-up Activities ········· 025
    PART TWO   Reading Activities ········· 028
        Passage 1   Textile Fibers ········· 028
        Passage 2   Natural Fibers ········· 036
        Passage 3   Synthetic Fibers and Blends ········· 042

Unit 3   Textile Yarns and Spinning Technology ········· 047
    PART ONE   Warm-up Activities ········· 047
    PART TWO   Reading Activities ········· 049
        Passage 1   Yarns and Their Classifications ········· 049
        Passage 2   Early History and Development of Spinning ········· 058
        Passage 3   Fiber Spinning ········· 063

Unit 4   Woven Fabrics and Weaving Technology ········· 067
    PART ONE   Warm-up Activities ········· 067
    PART TWO   Reading Activities ········· 070
        Passage 1   Recent Developments: Weaving Technology ········· 070
        Passage 2   Shuttleless Looms ········· 079
        Passage 3   Choosing Velvet 'Pile' Types of Fabric ········· 085

Unit 5   Knitting Technology ········· 091

	PART ONE	Warm-up Activities	091
	PART TWO	Reading Activities	094
	Passage 1	Modern Knitting Technology Trend: Seamless Technology	094
	Passage 2	The Origins of the Knitting Machine	100
	Passage 3	A More Casual Style of Men's Tie	105

Unit 6　Fashion Designers　111

	PART ONE	Warm-up Activities	111
	PART TWO	Reading Activities	114
	Passage 1	The Occupation: Fashion Designers	114
	Passage 2	Becoming a Fashion Designer	123
	Passage 3	Coco Chanel: The Legend and the Life	130

Unit 7　Fashion Marketing　139

	PART ONE	Warm-up Activities	139
	PART TWO	Reading Activities	141
	Passage 1	The New Era of Fashion Marketing	141
	Passage 2	Online Shopping: Fashion at Your Fingertips	148
	Passage 3	Zara and H&M: Fast Fashion on Demand	155

Unit 8　International Trade of Textile and Fashion　161

	PART ONE	Warm-up Activities	161
	PART TWO	Reading Activities	162
	Passage 1	Eastern Players in the Global Fibres Industry	162
	Passage 2	Weaving New Markets	170
	Passage 3	Markets and Value in Clothing and Modeling	177

Glossary　183

Acknowledgements　203

Unit 1　History of Textile and Fashion

PART ONE　Warm-up Activities

Silk Production——A State Secret

New Words

pupae /ˈpjuːpiː/ *n.* 蛹（复数）	cocoon /kəˈkuːn/ *n.* 茧
dissolve /dɪˈzɒlv/ *vt.* 使溶解、使分解	unravel /ʌnˈrævl/ *v.* 解开
pluck /plʌk/ *v.* 拨、扯	sticky /ˈstɪki/ *adj.* 黏的
coating /ˈkəʊtɪŋ/ *n.* 涂层、包衣	strand /strænd/ *n.* 线、串
bind /baɪnd/ *v.* 约束	unwind /ʌnˈwaɪnd/ *v.* 解开

Directions: *Listen to the recording and choose the most appropriate answer to each of the following questions.*

1. To which place did the two monks smuggle silkworm eggs?
   A. Constantinople.　　B. Constantine.　　C. Nepal.
2. How long can a single strand of silk be?
   A. Two to three thousand miles.
   B. Two to three thousand feet.
   C. Two or three thousand meters.
3. How many strands usually make the fine thread which is used to weave silk cloth?
   A. Four.　　　　　　B. Three.　　　　　C. Two.
4. Why should a cocoon be steamed?
   A. To kill the pupae inside.
   B. To dissolve the sticky coating that binds the silk.
   C. To make sure the strands will not be twisted.
5. When were silkworms first smuggled abroad?
   A. In 515 A.D.
   B. In 550 A.D.
   C. In 552 A.D.

American Textile History Museum (1)

New Words

associate /əˌsəuʃieit/ n. 同伴、合伙人	ingenuity /ˌindʒi'njuːiti/ n. 心灵手巧
micronize /'maikrənaiz/ vt. 使微粉化	demonstrate /'demənstreit/ v. 证明、演示
transformation /ˌtrænsfə'meiʃn/ n. 转化	ignite /ig'nait/ v. 引发、点燃、着火

Directions: *Listen to this part carefully and answer the following questions.*

1. Compared to traditional uses, what are some new development in textiles?
   _____

2. Where was the first textile plant established in America?
   _____

3. By whom were textiles initially produced in the United States?
   _____

4. For what purpose was the American Textile History Museum founded?
   _____

American Textile History Museum (2)

New Words

interrelated /ˌintəri'leitid/ adj. 相关联的	interacted /ˌintər'æktid/ adj. 相互作用的

Directions: *Listen to the passage and fill in each blank with the information you get from the recording.*

1. No other industry is more rooted in American history than textile manufacturing. Just as textiles have_____ our past, they will continue to _____ our future.

2. From old textiles to space-aged textiles, to the _____ to make them, this museum _____ the best _____ in the world.

3. I think one of the_____ of the museum is to get people excited and interested in textiles.

4. There's art you have the designers and the people who choose the colors. You have scientists who make the pigments. And all of these are _____ and make the final product.

5. The American Textile History Museum _____ more than 8500 students every year. At the museum, textile art, science, and history _____ ____ as students go hands-on with _____ exhibits and increase their understanding of the many ways textiles impact us.

6. The American Textile History Museum, stories of the past, _____ of the future.

PART TWO Reading Activities

Passage 1 History of Fabric and Textile

1 Textiles are defined as the yarns that are woven or knitted to make fabrics. The use of textiles links the myriad cultures of the world and defines the way they clothe themselves, adorn their surroundings and go about their lives. Textiles have been an integral part of human daily life for thousands of years. The first use of textiles, most likely felt, dates back to the late Stone Age, roughly 100,000 years ago. However, the earliest instances of cotton, silk and linen appeared around 5,000 BC in India, Egypt and China. The ancient methods of manufacturing textiles, namely plain weave, satin weave and twill, have changed very little over the centuries. Modern manufacturing speed and capacity, however, have increased the rate of production to levels unthinkable even 200 years ago.

Late antique textile, Egyptian, now in the Dumbarton Oaks collection.

History

2 Trade of textiles in the ancient world occurred predominantly on the Silk Road, a winding route across lower Asia that connected the Mediterranean lands with the Far East. Spanning over 5,000 miles and established during the Han Dynasty in China around 114 BC, the Silk Road was an integral part of the sharing of manufactured goods, cultures and philosophies, and helped develop the great civilizations of the world. During the Middle Ages, simple clothing was favored by the majority of people, while finer materials such as silks and linens were the trappings of royalty and the rich. During the 14th century, however, advances in dyeing and tailoring accelerated the spread of fashion throughout Western Europe, and drastically altered the mindset of both wealthy man and commoner alike. Clothing and draperies became increasingly elaborate over the next several centuries, although production methods remained largely unchanged until the invention of steam-powered mechanized facilities during the Industrial Revolution. From that point on, quality textiles became available to the masses at affordable prices.

Sources and types

3 Textiles can be derived from several sources. Animals, plants and minerals are the traditional sources of materials, while petroleum-derived synthetic fibers were introduced in the mid-20th century. By far, animal textiles are the most prevalent in human society, and are commonly made from furs and hair. Silk, wool, and pashmina are all extremely popular animal textiles. Plant textiles,

the most common being cotton, can also be made from straw, grass and bamboo. Mineral textiles include glass fiber, metal fiber and asbestos. The recent introduction of synthetic textiles has greatly expanded the array of options available for fabric manufacturers, both in terms of garment versatility and usability. Polyester, spandex, nylon and acrylic are all widely-used synthetic fibres.

Uses

4　Textiles have an assortment of uses, the most common of which are for clothing and containers such as bags and baskets. In the household, they are used in carpeting, upholstered furnishings, window shades, towels, covering for tables, beds, and other flat surfaces, and in art. In the workplace, they are used in industrial and scientific processes such as filtering. Miscellaneous uses include flags, backpacks, tents, nets, cleaning devices such as handkerchiefs and rags, transportation devices such as balloons, kites, sails, and parachutes, in addition to strengthening in composite materials such as fibreglass and industrial geotextiles. Children can learn using textiles to make collages, sew, quilt, and create toys.

5　Textiles used for industrial purposes, and chosen for characteristics other than their appearance, are commonly referred to as technical textiles. Technical textiles include textile structures for automotive applications, medical textiles (e.g. implants), geotextiles (reinforcement of embankments), agrotextiles (textiles for crop protection), protective clothing (e.g. against heat and radiation for fire fighter clothing, against molten metals for welders, stab protection, and bullet proof vests).

Production methods

6　In addition to the multitude of textiles available for use, there are many different methods for creating fabrics from textiles.

7　Weaving is a textile production method which involves interlacing a set of longer threads (called the warp) with a set of crossing threads (called the weft). This is done on a frame or machine known as a loom, of which there are a number of types. Some weaving is still done by hand, but the vast majority is mechanised.

8　Knitting and crocheting involve interlacing loops of yarn, which are formed either on a knitting needle or on a crochet hook, together in a line. The two processes are different in that knitting has several active loops at one time on the knitting needle waiting to interlock with another loop, while crocheting never has more than one active loop on the needle.

9　Lace is made by interlocking threads together independently, using a backing and any of the methods described above, to create a fine fabric with open holes in the work. Lace can be made by either hand or machine.

10　Felting involves pressing a mat of fibres together, and working them together until they become tangled. A liquid, such as soapy water, is usually added to lubricate the fibres, and to open up the microscopic scales on strands of wool.

11   Nonwoven textiles are manufactured by the bonding of fibres to make fabric. Bonding may be thermal or mechanical, or adhesives can be used.

Treatments

12   Textiles are often dyed, with fabrics available in almost every colour. The dyeing process often requires several dozen gallons of water for each pound of clothing. Coloured designs in textiles can be created by weaving together fibres of different colours, adding coloured stitches to finished fabric, creating patterns by resisting dyeing methods, tying off areas of cloth and dyeing the rest, or drawing wax designs on cloth and dyeing in between them, or using various printing processes on finished fabric. Woodblock printing, still used in India and elsewhere today, is the oldest of these dating back to at least 220 B.C. in China. Textiles are also sometimes bleached, making the textile pale or white.

13   Textiles are sometimes finished by chemical processes to change their characteristics. In the 19th century and early 20th century starching was commonly used to make clothing more resistant to stains and wrinkles. Since the 1990s, with advances in technologies such as permanent press process, finishing agents have been used to strengthen fabrics and make them wrinkle free.

14   More so today than ever before, textiles receive a range of treatments before they reach the end-user. However, many of these finishes may also have detrimental effects on the end user. A number of disperse, acid and reactive dyes have been shown to be allergenic to sensitive individuals. Further to this, specific dyes within this group have also been shown to induce purpuric contact dermatitis.

(1088 words)

New Words

acid /'æsid/ *n.* ［化］酸，酸性物质；*adj.* 酸的，酸性的，酸味的
acrylic /ə'krilik/ *n.* 丙烯酸纤维，腈纶
adorn /ə'dɔːn/ *vt.* 装饰；使生色（+with）
agrotextiles /ˌægrəu'tekstailz/ *n.* 农用织物
allergenic /æləˈdʒenik/ *adj.* 引起过敏症的，导致过敏的
asbestos /æs'bestəs/ *n.* 石棉
assortment /ə'sɔːtmənt/ *n.* 各种各样
bleach /bliːtʃ/ *vt.&vi.* 使（颜色）变淡，变白；漂白，（使）晒白，褪色
clothe /kləuð/ *vt.* 给……穿衣，为……提供衣服；覆盖，使披上（+in）
collage /kəu'laːʒ/ *n.* 拼贴画，拼贴艺术；杂烩；收集品；收藏品
composite /'kɔmpəzit/ *adj.* 混合成的，综合成的，复合的
crocheting /'krəuʃeiiŋ/ *n.* 钩编，钩编工艺

detrimental /ˌdetrɪˈmentl/ adj. 有害的，不利的（+to）
dermatitis /ˌdəːməˈtaitis/ n. ［U］皮（肤）炎
disperse /disˈpəːs/ n. 分散剂；vt. 驱散，解散，疏散；传播，散发
drapery /ˈdreipəri/ n. （总称）布匹；［U］纺织品
drastically /ˈdræstikəli/ adv. 大大地，彻底；激烈地
dyeing /ˈdaiiŋ/ n. 染色，染色工艺
elaborate /iˈlæbərit/ adj. 精巧的，详尽的，复杂的
embankment /imˈbæŋkm(ə)nt/ n. ［U］筑堤；（河、海的）堤岸，（铁路的）路堤
end-user n. 最终使用者，消费者
fabric /ˈfæbrik/ n. 织物
felt /felt/ n. 毛毡，毡制品；vt. 把……制成毡；用毡覆盖；vi. 毡合，毡化
felting /ˈfeltiŋ/ n. 毡化
filtering /ˈfiltəriŋ/ n. 过滤，过滤作用
finish /ˈfiniʃ/ v. 后整理
gallon /ˈgælən/ n. 加仑（液量单位，1 美制加仑 =3.785 升，1 英制加仑 =4.546 升）
geotextile /ˈdʒiəutekstail/ n. 土工织物
implant /imˈplaːnt/ vt. 埋置；灌输，注入；种植，［医］移植
integral /ˈintigrəl/ adj. 构成整体所必需的；不可缺的（+to）
interlace /ˌintəˈleis/ vt. 使交织，使组合；vi. 交错，组合，穿插
interlock /ˌintəˈlɔk/ v. （使）连锁，（使）联结，（使）连扣
knitting /ˈnitiŋ/ n. ［U］编织；（总称）编织物
lace /leis/ n. ［U］花边，蕾丝，饰带；［C］鞋带；带子
　　　　　v. 穿带子于，用带系（+up）；用花边等装饰
lacing /ˈleisiŋ/ n. 花边织法；结带；镶边；饰带，花边
linen /ˈlinin/ n. ［U］亚麻布，亚麻线（纱）；亚麻布制品（如床单、桌巾、内衣等）
loom /luːm/ n. ［C］织布机；［U］织造术；vt. 在织布机上织
loop /luːp/ n. （线，铁丝等绕成的）圈，环
lubricate /ˈluːbrikeit/ vt. 使滑润，给……上润滑油
mat /mæt/ n. 丛，簇，团（+of）；地席，草席；垫子
mindset /ˈmaindset/ n. 心态；倾向、习惯
miscellaneous /misiˈleinjəs/ adj. 混杂的，五花八门的，各种各样的；多才多艺的
myriad /ˈmiriəd/ n. 无数，大量（+of）；adj. 无数的，大量的；各种各样都有的
nylon /ˈnailən/ n. ［U］尼龙，锦纶；（pl.）尼龙长袜
pale /peil/ adj. 苍白的，灰白的；（颜色）淡的
parachute /ˈpærəʃuːt/ n. 降落伞
pashmina /pæʃˈmiːnə/ n. 开司米亚羊毛

polyester /ˌpɔliˈestə/ n. ［化］聚酯；涤纶
predominantly /priˈdɔminəntli/ adv. 占主导地位地，占优势地，显著地
prevalent /ˈprevələnt/ adj. 流行的，盛行的，普遍的（+among/in）
spandex /ˈspænˌdeks/ n. （作腰带、泳衣用的）弹性人造纤维（织物），氨纶
starch /stɑːtʃ/ n. ［U］淀粉；淀粉类食物；［U］（浆衣服等用的）淀粉浆
stitch /stitʃ/ n. 一针，针脚，线迹；［C；U］针法；编结法；v. 缝，绣，编结（+up）
synthetic /sinˈθetik/ adj. 合成的，人造的，综合的
tangle /ˈtæŋgl/ vt.&vi. 纠结，乱成一团
tailoring /ˈteilərɪŋ/ n. 裁缝业，成衣业
textile /ˈtekstail/ n. 纺织品
twill /twil/ n. 斜纹织物；vt. 把……织成斜纹；adj. 斜纹织物的
unthinkable /ʌnˈθiŋkəbl/ adj. 难以想象的，不可思议的，难以置信的；不可能的
upholster /ʌpˈhəulstə/ vt. 为（沙发、椅子等）装上垫子（或套子、弹簧等）（+in/with）；用（挂毯、家具等）布置（房间）；装潢
usability /ˌjuːzəˈbiliti/ n. 可用；合用；可用性
versatility /ˌvəːsəˈtiliti/ n. ［U］多用途，多功能；多才多艺
vest /vest/ n. ［美］背心，马甲，防护背心
warp /wɔːp/ n. ［纺］［the+S］（棉布的）经线
weft /weft/ n. ［纺］［the+S］纬线，纬纱；织品；薄云层
wax /wæks/ n. ［U］蜡，蜂蜡，石蜡，蜡状物；adj. 蜡制的；vt. 给……上蜡
weave /wiːv/ vt. 织，编，编制
weaving /ˈwiːviŋ/ vt. 编，织
welder /ˈweldə/ n. 焊工
woodblock /ˈwudblɔk/ n. 木板；木块；［印］木版；木刻（画）
yarn /jɑːn/ n. 纱、线

Phrases and Expressions

crochet hook	钩针
interlocking thread	连锁线
nonwoven textiles	非织造织物
permanent press process	耐久压烫工艺
plain weave	平纹机织
printing processes	印花工艺
purpuric contact dermatitis	接触性皮炎
reactive dyes	活性染料

resist dyeing methods	防染染色法
satin weave	缎纹组织
soapy water	肥皂水
technical textiles	科技织物
woodblock printing	版画

Key Sentences

1. Textiles are defined as the yarns that are woven or knitted to make fabrics.
纺织品是指用纱线通过编织或者针织方法制成的织物。

2. Textiles can be derived from several sources. Animals, plants and minerals are the traditional sources of materials, while petroleum-derived synthetic fibers were introduced in the mid-20th century.
纺织品的材料来源可以有好几种。动物、植物以及矿物是传统的材料来源，而以石油为原料生产的合成纤维是在 20 世纪中期发明的。

3. Plant textiles, the most common being cotton, can also be made from straw, grass and bamboo. Mineral textiles include glass fiber, metal fiber and asbestos.
植物纺织品中最常见的棉，也可以由稻草、草和竹子制得。矿物纺织品包括玻璃纤维、金属纤维和石棉。

4. Polyester, spandex, nylon and acrylic are all widely-used synthetic fibres.
涤纶、氨纶、锦纶和腈纶都是广泛使用的人工合成纤维。

5. Technical textiles include textile structures for automotive applications, medical textiles (e.g. implants), geotextiles (reinforcement of embankments), agrotextiles (textiles for crop protection), protective clothing (e.g. against heat and radiation for fire fighter clothing, against molten metals for welders, stab protection, and bullet proof vests).
产业用纺织品包括各种织物结构，用于汽车、医用（如埋植剂）、土工织物（用于加固堤坝等）、农用织物（用于保户农作物）、防护衣（如消防服装的耐热及耐辐射性，焊接工人服装的抗熔金属的特性，穿刺保护服装、防弹背心）。

6. Weaving is a textile production method which involves interlacing a set of longer threads (called the warp) with a set of crossing threads (called the weft).
编织是一种纺织品生产方法，将一系列的纵线（称为经线）与一系列的横线（称为纬线）相互交叉织成。

7. Knitting and crocheting involve interlacing loops of yarn, which are formed either on a knitting needle or on a crochet hook, together in a line.
针织和钩编是利用针织用针或者钩针将纱线弯曲成线圈，再使线圈相互串套成排。

8. Lace is made by interlocking threads together independently, using a backing and any of the

methods described above, to create a fine fabric with open holes in the work.

花边是独立使用连锁线完成，使用一个后退织法或者其他上述的任何方法制造具有孔洞的精致织物。

9. Coloured designs in textiles can be created by weaving together fibres of different colours, adding coloured stitches to finished fabric, creating patterns by resisting dyeing methods, tying off areas of cloth and dyeing the rest, or drawing wax designs on cloth and dyeing in between them, or using various printing processes on finished fabric.

织物的色彩设计可以使用不同方法，如将不同颜色的纤维编织在一起，或者在后整理织物上添加彩色缝线，或者通过防染染色工艺产生彩色图案，或者将织物的一部分绑住将剩余部分染色，或者在织物上涂蜡之后将无蜡部分染色，或者是在后整理织物上使用各种印花工艺。

10. Textiles are sometimes finished by chemical processes to change their characteristics.

有时人们通过化学工艺对纺织品进行加工整理以改变它们的特性。

Notes

1. The Middle Ages（中世纪）is a period of European history from the 5th century to the 15th century. The Middle Ages follows the fall of the Roman Empire in 476 and precedes the Early Modern Era. It is the middle period of a three-period division of Western history: Classic, Medieval and Modern.

2. weaving 织物的编织方法：

平纹织物（Plain Weave）是采用平纹组织的织物。经纬纱每隔一根交织一次，交织点排列稠密，正反面没有区别。平纹织物结构紧密，质地坚牢，但手感硬。一般绣花产品采用该种织物，缩水率相对斜纹织物低，牢固度相对斜纹织物高。

 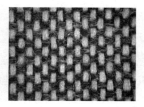 

face　　　　　　　　　　reverse

缎纹织物（Satin Weave）是采用缎纹组织的织物。缎纹织物的经纱或纬纱在织物中形成一些单独的、互不连接的经组织点或纬组织点，布面几乎全部由经纱或纬纱覆盖，表面似有斜线，但不像斜纹织物那样有明显的斜线纹路，经纬纱交织的次数更少，具有平滑光亮的外观，质地较柔软。

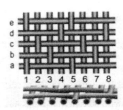 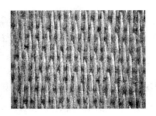 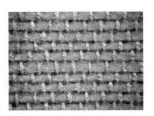

                                                face   reverse

  斜纹织物（Twill Weave）是采用斜纹组织的织物。布面有明显斜向纹路，手感柔软、有光泽，弹性较好，相对平纹织物缩水率较大。目前市面上印花产品多采用该种织物。

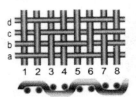  

                                                face   reverse

Post-Reading Exercises

- **Reading Comprehension**

Directions: *Read the passage and answer the following questions.*

1) According to the passage, what is the definition of "textile"?

2) When did the first use of textiles in human history appear?

3) What is "the Silk Road"?

4) How many sources are textiles derived from? What are they?

5) What are technical textiles?

6) How many different methods for creating fabrics from textiles? What are they?

7) How is the weaving done?

8) Which is the oldest kind of printing?
   _____

9) Why was starching commonly used to make clothing?
   _____

10) Give an example of a detrimental effect caused by some textile treatments.
    _____

- **Vocabulary**

Directions: *Read the following groups of sentences carefully and discuss with your partner how the same word is used with different meanings in each group. Then translate them into Chinese.*

1) clothe

    a. He has to work hard to feed and *clothe* his family.

    b. The trees are *clothed* in green leaves.

2) composite

    a. The play is a *composite* of reality and fiction.

    b. This *composite* illustration was made by putting together a number of drawings.

3) disperse

    a. The lips of the wise *disperse* knowledge.

    b. After school, the children *dispersed* to their homes.

    c. A novel *non-disperse* additive NDA for high performance underwater concrete was developed in this paper.

4) felt

    a. He was the consummate entertainer and his contributions and legacy will be *felt* by the world forever .

    b. He wears an old brown *felt* hat.

    c. It doesn't *felt* up after repeated washing.

5) finish

    a. The goal is to provide high school students a safe, drug-free environment where they can *finish* their education.

    b. It is the technical development of novel composite materials in terms of structure, production process, and *finish* on semi-products.

6) lace

    a. My room has white *lace* curtains at the enormous windows.

    b. Untie the *lace* and take off your shoe .

    c. Her fingers were too cold to *lace* the tent flap.

7) loop

a. He prefers hanging jeans by a belt *loop* to preserve their shape.

b. Ropes were being *looped* around him and he was helpless to resist.

c. Trails *loop* and weave through the tall trees.

8) warp

a. The shuttle moves backwards and forwards through the *warp*.

b. There is a *warp* in this record.

c. This wood *warps* easily in damp conditions.

d. Years of living alone may *warp* one's personality.

9) wax

a. The packet was then sealed with *wax*.

b. Get the car really clean and dry before you *wax* it.

10) versatility

a. He had impressed us with his *versatility* as a journalist.

b. Tofu's *versatility* lends it to preparation in a variety of ways including stewed, fried, grilled and raw.

- **Translation**

Directions: *Translate the following sentences into English, using the expressions in brackets.*

1）几千年来，纺织品是人类日常生活中必不可少的一部分。（an integral part of...）

2）经过几个世纪的发展，纺织品变得越来越精美。（elaborate）

3）人类对于纺织品的使用，可以追溯到石器时代，大约十万年前。（date back to）

4）本店销售的服装有各种各样的风格和色彩。（assortment）

5）纺织品可能有几种来源：动物、植物、矿物以及由石油衍生而来的人造纤维。（derive...from）

Passage 2 History of Silk

(Women striking and preparing silk, painting by Emperor Huizong of Song, early 12th century.)

1 According to Chinese tradition, the history of silk begins in the 27th century BCE. Its use was confined to China until the Silk Road opened at some point during the latter half of the first millennium BCE. China maintained its virtual monopoly over silk for another thousand years. Not confined to clothing, silk was also used for a number of other applications, including writing, and the color of silk worn was an important indicator of social class during the Tang Dynasty.

2 Silk cultivation spread to Japan in around 300 CE, and by 522 the Byzantines managed to obtain silkworm eggs and were able to begin silkworm cultivation. The Arabs also began to manufacture silk during this same time. As a result of the spread of sericulture, Chinese silk exports became less important, although they still maintained dominance over the luxury silk market. Silk production was brought to Western Europe, in particular to many Italian states, which saw an economic boom exporting silk to the rest of Europe. Changes in manufacturing techniques also began to take place during the Middle Ages, with devices such as the spinning wheel first appearing. During the 16th century France joined Italy in developing a successful silk trade, though the efforts of most other nations to develop a silk industry of their own were unsuccessful.

3 The Industrial Revolution changed much of Europe's silk industry. Due to innovations in spinning cotton, it became much cheaper to manufacture and therefore caused more expensive silk production to become less mainstream. New weaving technologies, however, increased the efficiency of production. Among these was the Jacquard loom, developed

The silkworm cocoon

for silk embroidery. An epidemic of several silkworm diseases caused production to fall, especially in France, where the industry never recovered. In the 20th century, Japan and China regained their earlier role in silk production, and China is now once again the world's largest producer of silk. The rise of new fabrics such as nylon reduced the prevalence of silk throughout the world, and silk is now once again a somewhat rare luxury goods, much less important than in its heyday.

4 The earliest evidence of silk was found at the sites of Yangshao culture in Xia County, Shanxi, where a silk cocoon was found cut in half by a sharp knife, dating back to between 4000 and 3000 BCE. The species was identified as bombyx mori, the domesticated silkworm. Fragments of primitive loom can also be seen from the sites of Hemudu culture in Yuyao, Zhejiang, dated to about 4000 BCE. Scraps of silk were found in a Liangzhu culture site at Qianshanyang in Huzhou, Zhejiang, dating back to 2700 BCE. Other fragments have been recovered from royal tombs in the Shang dynasty (1600—1046 BCE).

5 During the later epoch, the Chinese lost their secret to the Koreans, the Japanese, and later the Indians, as they discovered how to make silk. Allusions to the fabric in the Old Testament show that it was known in western Asia in biblical times. Scholars believe that starting in the 2nd century BCE the Chinese established a commercial network aimed at exporting silk to the West. Silk was used, for example, by the Persian court and its king, Darius III, when Alexander the Great conquered the empire. Even though silk spread rapidly across Eurasia, with the possible exception of Japan its production remained exclusively Chinese for three millennia.

6 The writings of Confucius and Chinese tradition recount that in the 27th century BCE a silk worm's cocoon fell into the tea cup of the empress Leizu. Wishing to extract it from her drink, the young girl of fourteen began to unroll the thread of the cocoon. She then had the idea to weave it. Having observed the life of the silk worm on the recommendation of her husband, the Yellow Emperor, she began to instruct her entourage in the art of raising silk worms-sericulture. From this point on, the girl became the goddess of silk in Chinese mythology. Silk would eventually leave China in the hair of a princess promised to a prince of Khotan. This probably occurred in the early 1st century CE. The princess, refusing to go without the fabric she loved, would finally break the imperial ban on silk worm exportation.

7 Though silk was exported to foreign countries in great amounts, sericulture remained a secret that the Chinese guarded carefully. Consequently, other peoples invented wildly varying accounts of the source of the incredible fabric. In classical antiquity, most Romans, great admirers of the cloth, were convinced that the Chinese took the fabric from tree leaves. This belief was affirmed by Seneca the Younger in his Phaedra and by Virgil in his Georgics. Notably, Pliny the Elder knew better. Speaking of the bombyx or silk moth, he wrote in his *Natural History* "They weave webs, like spiders, that become a luxurious clothing material for women, called silk."

(860 words)

New Words

allusion /ə'lu:ʒən/ *n.* 隐喻，典故；暗示，暗指
antiquity /æn'tikwəti/ *n.* 古代，古物，古迹
boom /bu:m/ *n.* 兴旺，繁荣
Byzantine /bai'zæntain/ *n.* 拜占庭人，拜占庭式建筑；
    adj. 拜占庭的，拜占庭式的；错综复杂的，诡计多端的
confine /kən'fain/ *vt.* 限制，紧闭，使局限；*n.* 范围，限制，约束
cultivation /ˌkʌlti'veiʃən/ *n.* 种植，栽培；教化，培养
dominance /'dɔminəns/ *n.* 支配，控制；统治，优势
embroidery /im'brɔidəri/ *n.* 刺绣，刺绣品
entourage /'ɔntura:ʒ/ *n.* 随行，随从人员；周围，环境
epidemic /ˌepi'demik/ *n.* 流行，蔓延；流行病，传染病；*adj.* 流行的，传染性的
epoch /'i:pɔk/ *n.* （新）时代，（新）时期；重要时期，值得纪念的事件（或日期）
Eurasia /ju'reiʒə/ *n.* 欧亚大陆
extract /ik'strækt/ *vt.* 提取，选取，摘录；*n.* 精华，提取物，摘录
fragment /'frægmənt/ *n.* 碎片，碎屑；片段；*vt.&vi.* （使）成碎片，（使）分裂
georgic /'dʒɔ:dʒik/ *n.* 田园诗
innovation /ˌinəu'veiʃ(ə)n/ *n.* 创新，革新，改革
loom /lu:m/ *n.* 织布机
millennia /mi'leniə/ *n. (pl.)* millennium 的复数
millennium /mi'leniəm/ *n.* 一千年
monopoly /mə'nɔpəli/ *n.* 垄断，独占，控制
moth /mɔθ/ *n.* 飞蛾，蛾子
Persian /'pə:ʃən/ *adj.* 波斯的，波斯人的，波斯语的；*n.* 波斯人，波斯语
Phaedra /'fi:drə/ *n.* [希神] 菲德拉
prevalence /'prevələns/ *n.* 流行，盛行，普遍
primitive /'primətiv/ *adj.* 原始的，早期的；简单的，粗糙的，质朴的，自然的；
    n. 原始人，原始事物
recount /ri'kaunt/ *vt.* 详述，列举；*n.* 重新计算
scrap /skræp/ *n.* 碎片，小块；*adj.* 零碎的
sericulture /ˌseri'kʌltʃə/ *n.* 养蚕（业）

Phrases and Expressions

bombyx mori   家蚕

date back to	追溯到，始于（某时期）
in its heyday	在全盛期，鼎盛期
Jacquard loom	提花织机

Key Sentences

1. Not confined to clothing, silk was also used for a number of other applications, including writing, and the color of silk was an important indicator of social class during the Tang Dynasty.

不只服装，丝绸还用于诸多其他用途，包括书写。在唐朝，丝绸服饰的颜色是社会等级的一个重要象征。

2. As a result of the spread of sericulture, Chinese silk exports became less important, although they still maintained dominance over the luxury silk market.

由于养蚕业的传播，中国的丝绸出口已不那么重要，但仍占据着奢侈丝绸市场的主要地位。

3. The rise of new fabrics such as nylon reduced the prevalence of silk throughout the world, and silk is now once again a somewhat rare luxury good, much less important than in its heyday.

诸如尼龙之类的新织物的产生削减了丝绸在全世界的流行地位，如今丝绸又重新成为比较稀缺的奢侈品，但与它在全盛时期不可同日而语。

4. The earliest evidence of silk was found at the sites of Yangshao culture in Xia County, Shanxi, where a silk cocoon was found cut in half by a sharp knife, dating back to between 4000 and 3000 BCE.

最早的丝绸证据发现于山西夏县的仰韶文化遗址，在那里发现了一颗被利刃割掉一半的茧壳，可追溯到公元前 4000 年～公元前 3000 年。

5. Even though silk spread rapidly across Eurasia, with the possible exception of Japan its production remained exclusively Chinese for three millennia.

尽管丝绸迅速传遍了欧亚大陆（也许日本是个例外），丝绸的制作完全专属中国长达三千年。

6. The writings of Confucius and Chinese tradition recount that in the 27th century BCE a silk worm's cocoon fell into the tea cup of the empress Leizu. Wishing to extract it from her drink, the young girl of fourteen began to unroll the thread of the cocoon.

孔子以及中国的传统著作都详载了在公元前 27 世纪，一颗蚕茧（意外）掉进了嫘祖（黄帝妃子）的茶杯中。想要从茶中取出蚕茧，这位 14 岁的年轻姑娘开始抽取蚕丝。

7. Speaking of the *bombyx* or silk moth, he wrote in his *Natural History*, "They weave webs, like spiders, that become a luxurious clothing material for women, called silk."

说到家蚕或丝蚕，他在《自然的历史》中写道："它们像蜘蛛一样织网，织成了奢华的女性服装面料，被称为丝绸。"

Notes

1. BCE : (abbr.) Before the Christian (or Common) Era 公元前。
2. CE : (abbr.) the Christian (or Common) Era 公元。

Post-Reading Exercises

- **Reading Comprehension**

Directions: *Read the passage and decide whether the following statements are true (T) or false(F).*

1) _____ The use of silk was confined to China until the Silk Road opened during the latter half of the first millennium.

2) _____ The colour of silk worn was an important indicator of social class during the Song Dynasty.

3) _____ As a result of the spread of sericulture, Chinese silk exports became more important, and they still maintained dominance over the luxury silk market.

4) _____ Silk is now a somewhat rare luxury good, and China is once again the world's largest producer of silk.

5) _____ It was Leizu, the god of silk in Chinese mythology, who began to teach people how to raise silk worms.

6) _____ Subsequently the Koreans, the Japanese, and the Indians also discovered how to make silk since the Chinese lost their secret.

- **Vocabulary**

Directions: *Complete the following sentences with the words and phrases given in the box.*

| allusion | being confined | epidemic | epoch |
| extract | monopoly | primitive | virtual |

1) In a _____ space, students can "visit" ancient battlefields, or "talk" with Shakespeare.

2) The government has a _____ on oil production in that country.

3) Women should feel free to pursue happiness and live as they choose, instead of _____ _____ by traditional marriage values.

4) Within days, a terrible _____ swept through the district. In less than a week, over half of the people there were dead.

5) Albert Einstein's Theory of Relativity marked a new _____ in history.

6) I read a brief _____ of his new novel on the subway and it has rather whetted my appetite

for it.

7) In _____ times, nature appeared to be a mysterious and formidable force and men found themselves powerless in the face of it.

8) Without naming names, the teacher criticized the students by _____.

- **Translation**

Directions: *Translate the following Chinese terms into English.*

1）丝绸之路 _____
2）丝绸刺绣 _____
3）出口丝绸至欧洲 _____
4）丝制品 _____
5）养蚕业 _____
6）由于养蚕业的传播 _____
7）抽取蚕丝 _____
8）开发丝绸贸易 _____

Passage 3 History of Jeans

1 Jeans were invented a little over a century ago; jeans are the world's most popular, versatile garment, crossing boundaries of class, age and nationality. From their origins as pure workwear, they have spread through every level of the fashion spectrum, embraced internationally for their unmatched comfort and appeal. Constantly in demand, they have survived the passing of both trends and time, capturing the ethos of each succeeding decade. While their charisma springs from their legendary American roots, their commercial strength rests on innovation and interpretation in the hands of jeanswear makers around the world.

2 In the mid 40s, the Second World War came to an end, and denim blue jeans, previously worn almost exclusively as workwear, gained new status in the U.S. and Europe. Rugged but relaxed, they stood for freedom and a bright future. Sported by men, women, and sharp teenagers, they seemed as clean and strong as the people who choose to wear them. In Europe, surplus Levi's were left behind by American armed forces and were available in limited supplies. It was the population's first introduction to the denim legend. Workwear manufacturers tried to copy the U.S. originals, but those in the know insisted on the real thing.

3 In the 1950s, Europe was exposed to a daring new style in music and movies and jeans took on an aura of sex and rebellion. When Elvis Presley sang in "Jailhouse Rock," his denim prison uniform carried a potent virile image. Girls swooned and guys were quick to copy the King. In movies like "The Wild One" and "Rebel Without a Cause," cult figures Marlon Brando and James Dean portrayed tough anti-heroes in jeans and T-shirt. Adults spurned the look; teenagers, even those who only wanted to look like rebels, embraced it.

4 By the beginning of the 60s, slim jeans became a leisurewear staple, as teens began to have real fun, forgetting the almost desperate energy of the previous decade, cocooned in wealth and security. But the seeds of change had been sown, and by the mid 60s jeans had acquired yet another social connotation—as the uniform of the budding social and sexual revolution. Jeans were the great equalizer, the perfect all-purpose garment for the classless society sought by the hippy generation. In the fight for civil rights, at anti-war demonstrations on the streets of Paris, at sit-ins and love-ins

everywhere, the battle cry was heard above a sea of blue.

5 Bell-bottoms hit their peak and creativity flourished in the 1970s. Customized denim—embroidered, studded and patched—became all the rage in fashionable St. Tropez, giving jeans a new glamorous profile. Gradually, the outward symbol of the alternative culture was integrated into mainstream society. Even "respectable" adults accepted denim in their wardrobe. The jeans culture had become associated with youth, and everybody wanted to remain young. Disco reigned, and denim dressed up for night. The ultimate sign of the appropriation of denim by the establishment was the designer jeans wave, which swept America just as the decade came to a close.

6 Designer jeans took hold in Europe, a sign of the rejection of the utopian ideals of the 70s and a return to affluence and status. A backlash surfaced in the form of "destroyed" denim, meant as the ultimate in anti-fashion but instantly a major trend. Riding the extremes of boom and bust, labels flooded the market, and then retrenched, as consumers got weary. Acid wash debuted in 1986 and revitalized the scene. In the midst of it all, Levi's launched its "back to basics" campaign.

7 The high living and conspicuous consumption of the 80s proved to many to be an empty pursuit, and the beginning of the 90s saw a widespread reevaluation of priorities. Facing the next millennium, people became more concerned with the environment, family life and old-fashioned values. This search for quality and authenticity helped to perpetuate the basics boom of the late 80s, leading to an interest in period originals and in newer lines that recaptured the details and fabrics of the past. Once again adapting to the spirit of the times, jeans represented an old friend, practical and modern yet linked to the purer, simpler life of days gone by.

(708 words)

New Words

affluence /ˈæfluəns/ n. 富裕，富足，丰富
authenticity /ˌɔːθenˈtisəti/ n. 确实，真实性
backlash /ˈbæklæʃ/ n. 反击，后冲；激烈反应，强烈反响
budding /ˈbʌdiŋ/ adj. 萌芽的，开始发育（发展）的，初露头角的；
　　　　　　　　 n. 发芽，萌芽；vt.&vi. 动词 bud 的现在分词
bust /bʌst/ n. 失败，破产，经济萧条
charisma /kəˈrizmə/ n. 超凡魅力，领袖气质
cocoon /kəˈkuːn/ vt. 包围，包裹；vi. 作茧；n. 茧，茧状物
connotation /ˌkɔnəˈteiʃ(ə)n/ n. 含义，言外之意，内涵
conspicuous /kənˈspikjuəs/ adj. 显著的，明显的，突出的
cult /kʌlt/ n. 狂热崇拜，迷信（对象）；异教，邪教
debut /ˈdeibjuː/ n. 首次露面，初次登场；vt.&vi. 首次演出

embrace /im'breis/ vi. 拥抱；vt. 拥抱，接受；包含，包围
embroider /im'brɔidə/ vt. 在……上刺绣，给……修饰；vi. 刺绣，修饰，镶边
equalizer /'i:kwəlaizə/ n. 均衡器，补偿器，平衡杆
ethos /'i:θɔs/ n. 民族精神，时代思潮，社会风气，气质
garment /'ga:mənt/ n. 衣服，服装，服饰；vt. 给……穿衣服
glamorous /'glæmərəs/ adj. 富有魅力的，迷人的
patch /pætʃ/ vt. 修补，补缀，拼凑；n. 补丁，碎片
perpetuate /pə'petʃueit/ vt. 使永存，保存
potent /'pəut(ə)nt/ adj. 强有力的，有效的
priority /prai'ɔrəti/ n. 重点，优先（权）
profile /'prəufail/ n. 简介，概况；外形，轮廓；vt. 描绘……的轮廓，为……写传略
rage /reidʒ/ n. 风行，狂热；狂怒，盛怒；vi. 盛行，流行；狂怒，大怒
recapture /ri:'kæptʃə/ vt. 重新获得，夺回，收复；n. 重占，夺回，收复
reign /rein/ vi. 盛行，支配，统治；n. 盛行，支配；君王统治，在位期
retrench /ri'trentʃ/ vt.&vi. 减少，削减（经费，开支等）
revitalize /ri:'vaitəlaiz/ vt. 使振兴，使复兴，使恢复
rugged /'rʌgid/ adj. 结实的，坚固的
spectrum /'spektrəm/ n. 范围，幅度，系列，光谱
spurn /spə:n/ vt.&vi. 摒弃，拒绝，貌视
staple /'steip(ə)l/ n. 主要产品，重要特色；订书钉，U 形钉；
                  adj. 最基本的，最重要的；vt. 用订书机装订
stud /stʌd/ vt. 镶嵌，点缀；n. 大头钉，饰钉，金属扣
surplus /'sə:pləs/ adj. 剩余的，多余的；n. 剩余，多余
swoon /swu:n/ vi. 晕厥，昏倒；心醉神迷，神魂颠倒；n. 晕厥，昏倒，狂喜
versatile /'və:sətail/ adj. 通用的，多功能的，多才多艺的
virile /'virail/ adj. 男性的，有男子气概的，强壮的
wardrobe /'wɔ:drəub/ n. 衣橱，衣柜；全部服装
weary /'wiəri/ adj. 疲倦的，厌烦的；vt. 使疲倦，使厌烦；vi. 疲倦，厌烦

Phrases and Expressions

bell-bottoms	喇叭裤
designer jeans	名牌牛仔裤
dress up	盛装打扮；乔装，伪装
in the know	知道内情的
leave behind	遗留，留下；追过，超过

love-in	（嬉皮士等的）谈情说爱的集会
old-fashioned	旧式的，老式的；过时的，守旧的
rest on	依靠，以……为基础
sit-in	静坐抗议，静坐罢工
spring from	来自，源于，出于

Key Sentences

1. Constantly in demand, they have survived the passing of both trends and time, capturing the ethos of each succeeding decade.

牛仔装一直深受欢迎，它们经历了潮流和岁月的更迭，引领了每一个年代的气质风貌。

2. When Elvis Presley sang in "Jailhouse Rock," his denim prison uniform carried a potent virile image. Girls swooned and guys were quick to copy the King.

当猫王埃尔维斯·普雷斯利在演唱《监狱摇滚》时，他的牛仔监狱制服传递了一个阳刚的男子汉形象。女孩们为之倾倒，男孩们争相效仿这位天王。

3. By the beginning of the 60s, slim jeans became a leisurewear staple, as teens began to have real fun, forgetting the almost desperate energy of the previous decade, cocooned in wealth and security.

到 20 世纪 60 年代初，紧身牛仔裤成了休闲装的主流款。那时青少年们开始享受真正的快乐，国泰民安，忘却了几乎让人绝望的上个年代。

4. Bell-bottoms hit their peak and creativity flourished in the 1970s. Customized denim—embroidered, studded and patched—became all the rage in fashionable St. Tropez, giving jeans a new glamorous profile.

20 世纪 70 年代，喇叭裤最流行，具有最活跃的创造力。各种定制的牛仔裤——绣花的，镶嵌的，补丁的——在时尚之都圣特罗佩斯风行一时，赋予了牛仔裤充满魅力的新形象。

5. The high living and conspicuous consumption of the 80s proved to many to be an empty pursuit, and the beginning of the 90s saw a widespread reevaluation of priorities.

20 世纪 80 年代明显的高消费对很多人来说是一个空洞的追求。20 世纪 90 年代伊始，许多观念定位被重新评估。

Notes

St. Tropez : A town of southeast France on the Mediterranean coast of the French Riviera. It is a noted seaside resort where many rich and famous people go on vacation.

Post-Reading Exercises

• **Reading Comprehension**

Directions: *Read the passage and decide whether the following statements are true(T) or false(F).*

1) _____ The commercial strength of jeans leans on innovation and interpretation in the hands of jeanswear makers around the world.

2) _____ When the First World War came to an end, denim blue jeans, previously worn almost exclusively as workwear, gained new status in the U.S.

3) _____ Girls swooned and guys were quick to copy the King because his denim prison uniform carried a a strong masculine image.

4) _____ By the beginning of the 60s, Bell-bottoms hit their peak and became a leisurewear staple.

5) _____ Facing the new millennium, people became more concerned with the environment, family life and up-to-date values.

• **Vocabulary**

Directions: *Complete the following sentences with the words given in the box.*

authenticity	conspicuous	millennium	glamorous
priority	perpetuate	swooned	versatile

1) The young girls _____ when they saw their favorite pop singer.

2) The life of a professional footballer is an exciting one, which attracts media attention and _____ women.

3) Lincoln is a _____ example of a poor boy who succeeded.

4) The Government gave top _____ to reforming the legal system.

5) The London Eye, erected in 1999, also known as the _____ Wheel is the largest Ferris wheel in Europe.

6) The _____ of the manuscript is beyond doubt.

7) They decided to _____ the memory of their leader by erecting a statue.

8) He's a very _____ performer; he can act, sing, dance, and play the piano.

• **Translation**

Directions: *Translate the following Chinese terms into English.*

1）跨越阶级、年龄和国籍的界限 _____

2）无与伦比的舒适性和吸引力 _____

3）性感和叛逆的气息 _____
4）定制的牛仔裤 _____
5）富有魅力的新形象 _____
6）主流趋势 _____
7）涌入市场 _____
8）繁荣与萧条 _____

Unit 2　Textile Fibers

PART ONE　Warm-up Activities

Silk

New Words

fiber /ˈfaibə/ n.　纤维	high-strength n.　高强度、高张力
replicate /ˈreplikeit/ vt.　模拟	silk glands n.　丝腺
hydrophilic /haidrəˈfilik/ adj.　（化）亲水的，吸水的	gel-like adj.　胶体状的
crystallize /ˈkristəlaiz/ vi.　结晶，具体化	soluble /ˈsɔljubl/ adj.　溶解的
hydrophobic /haidrəˈfəubik/ adj.　疏水的，恐水的	
spin /spin/ v.　纺纱，纺织；（蚕）吐丝（spun/span; spun; spinning）	

Directions: *Listen to the recording carefully and choose the most appropriate answer to each of the following questions.*

1. Which is not the reason why scientists long to understand how exactly the silkworm manages the process of silk weaving?

    A. Because they want to replicate the process.

    B. Because they want to produce high-strength and high-performance materials for sports and law enforcement.

    C. Because silk is a very useful natural fiber in the world.

2. Which of the following is the key to making silk?

    A. Animals' carefulness.

    B. The water content in its silk glands.

    C. The proteins.

3. Where does the strength of silk come from?

    A. Proteins.　　　　　B. Water.　　　　　C. Crystallization.

4. For what reason would the premature crystallization likely prove fatal to the little creatures?

    A. It would clog up their silk glands.

    B. The question has not yet been solved.

C. It would convert the proteins into silk threads.

5. What is not happening as the water decreases?

   A. The water-loving parts of the proteins fold together in chains.

   B. The folded chains of silk push together more and more to form larger and larger gel-like structures.

   C. The water-fearing parts of the proteins retain enough water.

Directions: *Listen to the passage again and decide whether the following statements are true or false. Put "T" for True and "F" for False in the spaces provided.*

____ 6. Scientists long to understand just how exactly the silkworm manages the process of silk weaving, and now they've got a thorough knowledge of it.

____ 7. Scientists know that the key to making silk lies in the animal's excellent management of the water content in its silk glands.

____ 8. According to the research, parts of the proteins in silk are hydrophilic, while other parts are hydrophobic.

____ 9. Silkworms manage to convert these two (hydrophilic and hydrophobic) proteins into silk threads without the proteins crystallizing before they are ready to spin them.

____ 10. The answer lies in the silkworm's ability to slowly decrease the water content in its silk glands.

New Fiber

New Words

beep /biːp/ v. 嘟嘟作响	monitor /ˈmɒnɪtə(r)/ n. 监测器
fluorine /ˈflɔːriːn/ n. 氟	molecular /məˈlekjələ(r)/ adj. 分子的，由分子组成的
blood flow 血流（量）	lopsided /ˌlɒpˈsaɪdɪd/ adj. 不对称的
imperceptible /ˌɪmpəˈseptəbl/ adj. 感觉不到的	
piezoelectric /paɪiːzəʊɪˈlektrɪk/ adj. 压电的	asymmetry /eɪˈsɪmətri/ n. 不对称

Directions: *Listen to this part carefully and answer the following questions.*

1. What have M.I.T. researchers developed to allow clothing to pick up sound?

   _____

2. What is the big key to the fiber?

   _____

3. What could make a shirt a 24-hour blood-pressure monitor?

   _____

Directions: *Listen to the passage again and fill in each blank with the information you get from the recording.*

4. _____ is so yesterday. I mean, really, what can it do? It can't pick up sound, or 5. _____ _____ us if something's wrong. Or can it? M.I.T. researchers say they've developed a fiber that would allow clothing to eventually do those things. Their study is in the journal Nature Materials.

A big key to the fiber is a plastic commonly used in microphones. Its particular molecular structure involves a lopsided arrangement of fluorine atoms on one side and hydrogen atoms on the other. That asymmetry makes the plastic piezoelectric: it changes shape when it 6. _____ an electric field.

So any electric current will then cause the fibers to 7. _____. The fibers could act as a microphone or a speaker, depending on whether the vibrations were being recorded or amplified.

The clothing mic could capture speech, sure. But it could also monitor health by detecting almost 8. _____ sounds from the body. Sounds are like blood flow, which could make a shirt a 24-hour blood-pressure monitor. So maybe someday, your clothes will say: "I detect a rise in blood pressure. Please sit down."

PART TWO Reading Activities

Passage 1 Textile Fibers

1 The term "textile" is often used as another word for clothes, but in reality textile refers to any object produced from fibers. Items as diverse as carpets, T-shirts, parachutes, hats, ropes, bullet-proof vests, bags, tents and lingerie are all products of the textile industry. Hence, a study of textiles unifies fields of study as diverse as climatology, chemistry, physics, sociology, economics, and, of course, art and design.

2 Textile products surround each of us at every moment of our lives, from birth to death, waking or sleeping. Each day every person makes decisions about textiles. From the simplest choice of what clothes to wear to the commitment to buy a new quality blanket, judgements about the performance, durability, attractiveness, and care of textiles are consciously or unconsciously made.

3 Fibers are the foundation of the textile industry. All textiles are made up of fibers. All of the production flows and formulae in textile wet processing stages including pre-treatment, dyeing, printing and final finishing are designed and conducted on the basis of the properties of the fibers from which the textile are made.

4 What is a fiber? Some characteristics can be identified which all textile fibers must have if they are to be commercially successful: a high length to diameter ratio, strength, extensibility and elasticity, resistance to chemicals, heat, and sunlight, and ability to take color.

Length to diameter ratio

5 Fibers generally have a small cross-sectional area and a length that greatly exceeds the diameter. For cotton and wool, the length to diameter ratio is in the region of 2000 : 1 to 5000 : 1. These fibers are produced naturally in short lengths, known as staple fiber. The fiber lengths vary from 10 to 50 millimeters for cotton and from 50 to 200 millimeters for wool. Man-made fibers can be produced with many kilometers of yarn on a single package. The length to diameter ratio of the fiber is then infinite. This type of fiber is termed continuous filament. Silk is the only natural continuous filament fiber. Many man-made fibers are also produced as staple, so that they can be processed on the same machinery as the natural staple fibers.

Strength

6 The strength of a textile material ultimately depends on the strength of the individual fibers from which it is made. Consequently, fibers must have a certain level of strength if they are to be useful. A high strength is clearly more important in fibers used for reinforcement of the rubber in a

tire than for the fibers used in a knitted jumper.

Extensibility and elasticity

7 In use, stresses will frequently be applied to textile materials. The materials need to extend under the stress and be flexible. The fibers in a pair of tights need to extend every time when the wearer bends her legs. But having extended, the fibers need to be elastic and return to their original length. If they do not, the tights will quickly become wrinkly at the knees and ankle. Tights are just one particular application. All fibers need to be extensible and elastic, but to different degree.

Resistance to chemicals, heat and sunlight

8 In normal use and during care procedures, fibers will be exposed to conditions that may damage them. These conditions may include chemicals such as acids, alkalis, bleaches, detergents, or organic solvents including dry cleaning fluid, and physical effects such as heat or sunlight. The extent to which fibers are exposed depends on the particular use. Resistance to sunlight is more important in curtains than it is in underwear. Almost all fibers are exposed to harmful conditions to some extent. Domestic washing powders are mildly alkali and contain bleaches. The temperature during normal ironing can easily reach 200℃. The effect of the high temperature on the fibers will be slow in most cases and involves some weakening, with perhaps yellowing of white fibers and loss of brightness of colored products.

Ability to be colored

9 Most fibers are normally an off-white color. Life would be very dull if all textile products were off-white. Consequently, fibers need to be colored. Ideally, they should be colored by dyeing at a late stage of processing. This enables a quick response to customers' demands for the latest shade.

10 Fibers are usually grouped in order to research or discuss or apply them conveniently. Based on their chemical composition, fibers can be classified into many groups such as cellulosic fiber, protein fiber, viscose fiber, polyamide fiber, polyester fiber, polyacrylic fiber, etc. But the most convenient grouping divides them into two basic groups according to their origins: i.e. natural and man-made fibers. Natural fibers refer to all fibers that occur in fiber form in nature, including cotton, linen, wool, silk, and so on, which have been known and used for thousands of years. As natural fibers cannot meet the requirement of people, many polymers that do not naturally exist in the form of fiber have been processed into fiber form, usually by forcing the viscous polymers through a spinneret that consists of a series of tiny holes arranged in a circle, and used as fibers. These products are known as man-made fibers. Most of the man-made fibers have only been produced in the last 40 years, but they have made a great difference to present-day society, in the types of clothes that we wear as well as the comfort and convenience of living.

11 The two basic groups can then be further subdivided. The natural fibers can be subdivided into the three types of cellulosic, protein and mineral fibers according to their origins. The cellulosic fibers come from plant materials, the protein fibers come from animal sources, and there is a

mineral fiber in nature, which is asbestos. Man-made fibers are usually subdivided into two groups: regenerated and synthetic fibers. Regenerated man-made fibers are made from natural materials that cannot be used for textiles in their original form, but that can be regenerated into usable fibers by chemical treatment and processing. These regenerated fibers are made from such diverse substances as wood, corn protein, small cotton bits called linters, and seaweed. True synthetic man-made fibers are made or "synthesized" completely from chemical substances such as petroleum derivatives. And it was not until 1885, when the first man-made fiber, rayon, was produced commercially, that man began to make use of both natural fiber and man-made fiber to produce textile products.

(1069 words)

New Words

acid /'æsid/ *n.* 酸；*adj.* 酸的，酸性的
alkali /'ælkəlai/ *n.* 碱；*adj.* 碱性的
asbestos /æz'bestəs/ *n.* 石棉；*adj.* 石棉的
bleach /'bli:tʃ/ *n.* 漂白剂
brightness /'braitnis/ *n.* 明亮，鲜艳，鲜艳度，（色彩）明度
climatology /klaimə'tɔlədʒi/ *n.* 气候学，风土学
detergent /di'tə:dʒənt/ *n.* 洗涤剂，净洗剂
durability /djuərə'biləti/ *n.* 耐久性，持久性
dyeing /'daiiŋ/ *n.* 染色；*adj.* 染色的
elasticity /elæs'tisəti/ *n.* 弹性，弹性学，弹力，伸缩力
extensibility /ikstensə'biləti/ *n.* 伸长性，延伸性，展开性
fabric /'fæbrik/ *n.* 织物，织品，布
filament /'filəmənt/ *n.* 丝，长丝，细丝
finishing /'finiʃiŋ/ *n.* 后整理，织物整理
ironing /'aiəniŋ/ *n.* 熨烫
jumper /'dʒʌmpə/ *n.* 妇女穿的套头外衣，连兜头帽的皮外衣，套头衫
knitted /'nitid/ *adj.* 针织的，编织的
polyacrylic /'pɔljəkrilik/ *n.* 聚丙烯酸化合物，腈纶
polyamide /pɔli'æmaid/ *n.* 聚酰胺（尼龙）
polyester /pɔli'estə/ *n.* 聚酯，涤纶
polymer /'pɔlimə/ *n.* 聚合物
printing /'printiŋ/ *n.* 印花，印花工艺，印刷
protein /'prəuti:n/ *n.* 蛋白质；*adj.* 蛋白质的
resistance /ri'zistəns/ *n.* 阻抗性，抵抗，抵抗力，阻力，电阻

shade /ʃeid/ n. 颜色，色调，色泽，色光
spinneret /'spinərət/ n. 纺丝头，喷丝头
staple /'steipl/ n. 纤维，短纤维，毛束，纤维长度；订书钉
strength /streŋθ/ n. 强力，强度
textile /'tekstail, -til/ n. 纺织品，织物
viscose /'viskəus/ n. 黏胶液，黏胶纤维
wrinkle /'riŋkl/ n. 皱纹，褶皱；v. 起皱

Phrases and Expressions

cellulosic fiber/ cellulose fiber	纤维素纤维
crochet hook	钩针
diameter ratio	直径比，内外径比
length to diameter ratio	长径比
organic solvent	有机溶剂
pre-treatment	前处理，预处理
regenerated fiber	再生纤维
synthetic fiber	合成纤维
wet processing	湿加工

Key Sentences

1. Some characteristics can be identified which all textile fibers must have if they are to be commercially successful: a high length to diameter ratio, strength, extensibility and elasticity, resistance to chemicals, heat, and sunlight, and ability to take color.

成功的商品化的纺织纤维必须具有以下特性：高的长径比、强度、延伸性和弹性、耐化学药品、耐热和耐日晒以及着色性。

2. The strength of a textile material ultimately depends on the strength of the individual fibers from which it is made.

纺织材料的强度根本上取决于构成它的单纤维的强度。

3. In normal use and during care procedures, fibers will be exposed to conditions that may damage them.

在正常的使用和护理过程中，纤维暴露在外会受到损坏。

4. The temperature during normal ironing can easily reach 200℃.

在正常熨烫过程中温度很容易达到200℃。

5. The effect of the high temperature on the fibers will be slow in most cases and involves some

weakening, with perhaps yellowing of white fibers and loss of brightness of colored products.

在大多数情况下，高温对纤维的影响是缓慢的，包括使纤维的某些性能弱化，可能使白色的纤维泛黄和使有色产品的色彩鲜艳度降低。

6. Life would be very dull if all textile products were off-white.

如果所有的纺织品都是灰白色的，我们的生活会变得非常暗淡。

备注：虚拟语气，主句为 would+动词原形结构，从句为 if+主语+动词过去式结构，表示与现在的事实相反。

7. As natural fibers cannot meet the requirements of people, many polymers that do not naturally exist in the form of fiber have been processed into fiber form, usually by forcing the viscous polymers through a spinneret that consists of a series of tiny holes arranged in a circle, and used as fibers.

当天然纤维不能够满足人们的需求时，许多并不是以纤维形式天然存在的聚合物被加工成纤维形状，当作纤维使用；这个加工过程通常是将这些黏性聚合物挤压通过由一系列呈圆形排列的微孔所组成的纺丝头。

8. Most of the man-made fibers have only been produced in the last 40 years, but they have made a great difference to present-day society, in the types of clothes that we wear as well as the comfort and convenience of living.

大部分化学纤维在最近 40 年才被生产出来，但是它们使现代社会发生了很大的改变，如我们穿着的衣服类型、生活的舒适性和方便性。

9. And it was not until 1885, when the first man-made fiber, rayon, was produced commercially, that man began to make use of both natural fiber and man-made fiber to produce textile products.

直到 1885 年，第一种化学纤维黏胶投入商品生产后，人类才开始同时利用天然和化学两种纤维来生产纺织品。

备注：本句是强调句，It 称为强调句型引导词，本身无意义，只起引导作用。强调句的结构是 It + be + 被强调部分（通常是主语、宾语或状语等）+ that (who) + 其他部分。

Notes

1. Lingerie

Lingerie 是英国本土最具影响力的女性内衣品牌之一。内衣以其细腻、考究的工艺，深得女士的喜爱。Lingerie 品牌为促销产品想出新招数，聘请了一批英国著名的女流行歌手担任内衣模特，这些歌手都具备天使面孔与魔鬼身材。如果说在巴黎高级时装界克里斯汀·迪奥堪称魅惑的代名词，那么 Lingerie 在内衣界中有着异曲同工之妙。

2. 纺织纤维

纤维是天然或人工合成的细丝状物质，纺织纤维则是指用来制造纺织品的纤维。纺织纤维具有一定的长度、细度、良好的弹性和强力等物理性能以及较好的化学稳定性。纺织纤维一般分为天然纤维和化学纤维。天然纤维包括植物纤维、动物纤维和矿物纤维。植物纤维有棉、麻等纤维；动物纤维有羊毛、蚕丝、兔毛等纤维；矿物纤维有石棉纤维。化学纤维主要包括再生纤维和合成纤维。其中，再生纤维有黏胶纤维、醋酯纤维等；合成纤维有锦纶、涤纶、腈纶、氨纶、维纶、丙纶等。

3. 短纤维

指长度在几毫米至几十毫米的纤维，如棉、毛、麻等天然纤维，也可以是由长丝切断后制成。

4. 长丝

指连续的纤维，如蚕丝及化纤制丝时喷出的连续丝束。

Post-Reading Exercises

- **Reading Comprehension**

Directions: *Read the passage carefully and answer the following questions.*

1) According to the passage, what is the definition of "textile"?

2) What is the foundation of the textile industry?

3) What characteristics must textile fibers have to be commercially successful?

4) Why is the length to diameter ratio of the fiber infinite?

5) What does the strength of a textile material ultimately depend on?

6) Why do fibers need to be extensible and elastic?

7) According to the text what will damage fibers?

8) Why should fibers be colored at a late stage of processing?

9) When were most man-made fibers produced?

10) What is the difference between regenerated and synthetic fibers?

- **Vocabulary**

Directions: *Read the following groups of sentences carefully and discuss with your partner how the same word is used with different or similar meanings in each group. Then translate them into Chinese.*

1) used

   a. My sister and brother *used* too many antibiotics.

   b. Everything that *used* to be a sin is now a disease.

   c. Natural materials cannot be *used* for textiles in their original form.

2) clothe

   a. He can earn enough to feed and *clothe* his family.

   b. Along the streets are trees *clothed* in green leaves.

   c. The man is *clothed* from head to foot in black.

3) wear

   a. I would never *wear* that ugly shirt.

   b. She *wears* the medal on a string round her neck.

   c. Apart from tears, only time could *wear* everything away.

4) type

   a. He *types* what he composes everyday.

   b. The doctor has *typed* her blood.

   c. He was *typed* as villain in the theater.

5) particular

   a. Our teacher is very *particular* when it comes to punctuation.

   b. Pay *particular* attention to the poet's choice of words.

   c. Is there any *particular* food you'd prefer?

6) bend

   a. This slender tree *bends* when the wind blows hard.

   b. The person who *bends* to such an unjust law must break his own moral code.

   c. She couldn't *bend* her mind to her studies.

7) slow

   a. The buyers were *slow* to act, and the house was sold to someone else.

   b. He was a quiet boy who seldom spoke, and some people thought he was a little *slow*.

   c. Business is *slow* during the summer.

8) mildly

   a. She was *mildly* bloated from head to toe.

   b. He put the case very *mildly*.

   c. To put it *mildly*, she is too poor in English.

9) staple

    a. Sweet potato was the *staple* of their diet.

    b. That is a positively healthy outlook, however, compared with another *staple* of American life: the home telephone.

    c. She seems to be the *staple* topic of conversation at the moment.

10) wrinkle

    a. This shirt will not *wrinkle*.

    b. She's beginning to get *wrinkles* round her eyes.

    c. He *wrinkled* up his forehead in perplexity.

- **Translation**

Directions: *Translate the following sentences into English, using the expressions in brackets.*

1）纤维是纺织工业的基础。（foundation）

2）棉花的纤维长度范围为 10 ～ 50 毫米。（vary from...to...）

3）根据其化学组成，纤维可以分成许多种类。（be classified into...）

4）这些产品被称为人造纤维。（be known as）

5）如果所有的纺织品都是灰白色的，我们的生活会变得非常暗淡乏味。（dull）

Passage 2 Natural Fibers

1 Fiber, the primary material from which most textile products are made, can be defined as units of hairlike dimensions, with a length at least one hundred times greater than the width. The textile fibers may be divided into two major groups according to their origins: (1) natural fibers, (2) chemical fibers.

2 Those that are found in nature are known as natural fibers, which are taken from either animal, vegetable or mineral resources. The classification of natural fibers is shown in Table 2-1. Vegetable fibers could be further divided according to the part of the plant that produces the fibers: the leaf, a hair produced from a seedpod, or the stem. The latter is called bast fibers. Animal fibers could be further divided into those fibers from the hair of an animal such as wool, and those from an extruded web such as silk. Chemically, the classification might be cellulosic for vegetable fibers, protein for animal fibers, and name of the specific minerals (such as asbestos) for mineral fibers. Using this scheme, cotton is a plant seedpod, or cellulosic fiber, and wool is an animal, hair, or protein fiber. They vary considerably as regard their properties and their production.

Table 2-1 Classification of Natural Fibers

Cellulosic fibers	Protein fibers	Mineral fibers
Cotton	Wool	Asbestos
Flax	Silk	
Jute	Mohair	
Ramie	Cashmere	
	Other animal hair	

Cotton

3 Cotton is by far the most important textile fiber and makes up nearly 50 percent of the total weight of fibers used in the world. It is obtained from the cotton plant which grows in warm, moist climates, and in most parts of the world. Cotton fibers are composed largely of cellulose. Besides cellulose, raw cotton contains a number of other substances, notably waxes, pectic products and mineral substances. They can range from 4% to 12% together and are referred to as impurities by the manufacturer of cotton goods. Generally these are objectionable effects and would make it difficult to color and finish cotton fabrics satisfactorily, so it is always a first step in the art of dyeing and finishing to purify the cotton as completely as possible. Cellulose is considered to be a condensation polymer formed from the glucose units. There are 3,000—5,000 glucose units joined together in natural cellulosic fibers. This corresponds to a molecular weight of 300,000—500,000.

4 Cotton is excellent for a multitude of purposes and has virtually universal consumer accep-

tance. It is used for apparel fibers, for household or domestic goods, and for industrial applications. Cotton is also extensively used in blends with man-made fibers to achieve new combinations of properties that are not available in the fibers separately.

5 Cotton has some disadvantages, too. It creases and wrinkles easily. It is readily attacked by acid reagents or substances, and it is slowly affected by sunlight, causing yellowing and fiber degradation.

Flax (Linen)

6 Flax is a bast fiber obtained from the flax plant. Flax fibers resemble cotton in so far as they consist of cellulose but have lower cellulose content. On an average the flax fibers contain only about 75 percent of pure cellulose, the remaining matter being a gummy pectic substance. The polymer of flax consists of a degree of polymerization of about 36,000 glucose units.

7 The surface of each fiber is smooth and this helps to give linen materials their characteristic high luster. Flax has relatively high strength. In many of its chemical properties linen closely resembles cotton. Thus, it is resistant to alkalis and is easily deteriorated by acids. Linen is mainly used in the manufacture of sail cloth, tent fabric, sewing threads, fishing lines, table-cloth and sheets, but today it is often used as a component of blends.

Wool

8 Wool is the most important fiber and produced in the largest amount. Wool is the fur-like covering of sheep that are raised in many countries around the world. It is obtained by shearing the fibrous covering from the sheep.

9 Wool fibers are composed of protein in which the repeated unit is amino acid. The amino acids are linked to each other by the peptide bond (—CO—NH—) to form the protein polymer. Chemically, the most important component in wool is keratin that is a complex protein and composed of 16 to 18 different amino acids. Keratin is amphoteric in nature. So wool can be dyed with acid or reactive dyes.

10 Most wool fibers have a white or creamy color, although some breeds of sheep yield brown or black wool. Wool fibers have a tendency to return completely to their original shape after small deformations, which is a great asset in apparel fabrics. The natural crimp in wool is of great importance, since it results in making a yarn fluffy, thereby trapping air in the interstices between the fibers. This trapping of air helps in forming an insulating layer, thus imparting the characteristic of warmth. Wool has several disadvantages: it is very sensitive to alkaline substances, it is readily attacked by moths and carpet beetles unless treated to resist them, it is difficult to bleach, and it felts easily.

11 According to their fineness and length, wool can be divided into four types: fine wool, long wool, medium wool and carpet wool. Wool is used primarily in apparel and home furnishings.

Silk

12 Silk is the material extruded from gland in the body of the silkworm in spinning its cocoon or web. To reclaim silk filaments, the cocoons are soaked in hot water, which soften the sericin gum. Filaments from several cocoons are picked up, assembled, passed through a guide, and made into skeins by the process of reeling. This yarn can then be processed into fabrics before or after degumming.

13 Raw silk is composed of two proteins: fibroin and sericin. Fibroin is the actual fiber protein. Sericin is the gummy substance and holds the filament together. The average composition of raw silk is 70% ~ 75% fibroin, 20% ~ 25% sericin, 2% ~ 3% waxy substances, and 1% ~ 1.7% mineral matter. Both fibroin and sericin are built up of 16 ~ 18 amino acids. The degree of polymerization of silk fibroin is uncertain, with DP of 300 to 3,000 having been measured in different solvents.

14 Silk fibers have a smooth surface and a distinctive triangular cross-sectional dimension, so silk fibers have a lustrous appearance. Silk has tenacity and relatively good moisture regain. Silk is warm and pleasant to touch and is generally considered comfortable to wear. It is readily dyeable with a variety of dyes. Silk is popular in men's neckties for its band and drape. The fiber is used alone and in blends with other fibers.

(1085 words)

New Words

apparel /ə'pærəl/ *n.* 服饰，服装
cashmere /'kæʃmiə/ *n.* 开司米，山羊绒
degradation /degrə'deiʃən/ *n.* 降解；降低
denier /di'niə/ *n.* 极少量；旦尼尔
elongation /i:lɔŋ'geiʃən/ *n.* 伸长
extension /ik'stenʃən/ *n.* 伸长
felt /felt/ *v.* 缩绒；*n.* 毛毡，毡制品
flax /flæks/ *n.* 亚麻
glucose /'glu:kəus/ *n.* 葡萄糖
guide /gaid/ *n.* 导丝器
hand /hænd/ *n.* 手感
jute /dʒu:t/ *n.* 黄麻
keratin /'kerətin/ *n.* 角蛋白
linen /'linin/ *n.* 亚麻，亚麻布，亚麻制品
luster /'lʌstə/ *n.* 光彩，光泽
mohair /'məuheə/ *n.* 马海毛，安哥拉羊毛

pectic /'pektik/ *adj.* 果胶的，黏胶质的
ramie /'ræmi/ *n.* 苎麻，苎麻纤维
reclaim /ri'kleim/ *v.* 缫丝，回收
reel /ri:l/ *v.* 缫丝
resiliency /ri'ziliənsi/ *n.* 弹性
sericin /'seərisin/ *n.* 丝胶
skein /'skein/ *n.* 绞纱，一束（线或纱）
spin /spin/ *v.* 纺纱，纺织
tenacity /ti'næsəti/ *n.* 韧性，强度

Phrases and Expressions

acid dyes	酸性染料
bast fiber	韧皮纤维
chemical fiber	化学纤维
cross-section	横截面
dimensional stability	尺寸稳定性
mineral fiber	矿物纤维
moisture absorption	吸湿性
moisture regain	回潮率
natural fiber	天然纤维
peptide bond	肽键
reactive dyes	活性染料

Key Sentences

1. Cotton is by far the most important textile fiber and makes up nearly 50 percent of the total weight of fibers used in the world.

棉花是最重要的纺织纤维，将近占世界所用的纤维总重量的 50%。

2. Generally these are objectionable effects and would make it difficult to color and finish cotton fabrics satisfactorily, so it is always a first step in the art of dyeing and finishing to purify the cotton as completely as possible.

一般来说，这些物质的存在都会有不良的影响，会使棉织物的染色和整理难以令人满意，因此，在染整工艺中，第一步总是尽可能地去除棉纤维的杂质。

3. The natural crimp in wool is of great importance, since it results in making a yarn fluffy, thereby trapping air in the interstices between the fibers. This trapping of air helps in forming an

insulating layer, thus imparting the characteristic of warmth.

羊毛的自然卷曲非常重要，因为这会导致纱线蓬松，从而使空气在纤维空隙之间滞留。这种空气滞留有助于形成绝缘层，从而赋予羊毛特有的保暖特性。

4. The degree of polymerization of silk fibroin is uncertain, with DP of 300 to 3,000 having been measured in different solvents.

丝素的聚合度是不确定的，在不同的溶剂中测定的聚合度为 300 ～ 3000。

Notes

1. 回潮率：纤维含水重量占纤维干重的百分比，即：回潮率＝（纤维湿重－纤维干重）/纤维干重。
2. 聚合度：指聚合物分子链中相同结构单元（或称链节）的个数，用 n 表示。

Post-Reading Exercises

• **Reading Comprehension**

Direction: *Read the passage carefully and complete the following sentences.*

1) Chemically, the classification of natural fibers might be _____ fibers, _____ fibers and _____ fibers.

2) Cotton fibers are composed of _____, _____, _____ _____ and _____ substances.

3) On an average, the flax fibers contain only about _____ of pure cellulose, the remaining matter being a gummy _____ substance.

4) _____ fibers are composed of _____ in which the repeated unit is _____ _____, which is linked to each other by the _____ _____ (—CO—NH—) to form the protein polymer.

5) Silk fibers have a lustrous appearance, because they have a _____ surface and a distinctive _____ _____ _____.

• **Vocabulary**

Direction: *Complete the following sentences with the words given in the box.*

felt	degradation	reel	reclaim
spin	resiliency	luster	tenacity

1) According to the tests we carried out, it's a non-active _____ product with no effect on the body.

2) The upper layer of_____ should overlap the lower.

3) Companies that _____ the plastic resin from empty beverage bottles say they can't get enough of the stuff.

4) Michelle will also _____ a customer's wool fleece to specification at a cost of $ 2.25 an ounce.

5) Compression _____ is one of the most important characteristics of carpets.

6) This is the method of test for breaking _____ of wool fibre bundles.

7) Mother asks me to help her to wind the wool around the _____ to prevent it from getting entangled（缠在一起的）.

8) These products have crystal _____ and strong vision effect.

- **Translation**

Direction: *Translate the following Chinese terms into English.*

1）天然纤维 _____
2）化学纤维 _____
3）韧皮纤维 _____
4）酸性染料 _____
5）活性染料 _____
6）肽键 _____
7）氨基酸 _____
8）横截面 _____

Passage 3 Synthetic Fibers and Blends

1 It took many years to develop the first spinning solution and to devise spinnerets to convert the solution into filaments. The first solutions were made by treating cellulose so it would dissolve in certain substances. It was not until the 1920s and 1930s that man first learned how to build long chain molecules from simple substances. The invention of nylon, and its successful marketing after the war, stimulated the synthesis of additional fibers. Gradually a wide variety of synthetic fibers were introduced for public consumption.

2 Melt spinning utilizes the thermoplastic characteristics of fibers. The spinneret holes are usually round, but noncircular holes are also used to make filaments of various cross-sectional shapes. The diameter of the fiber is determined by the rate at which the polymer is supplied to the hole in the spinneret and the wind-up speed, not by the diameter of the hole. For those polymers that are adversely affected by heat at or close to their melting temperatures so that they cannot be melt spun, dry spinning or wet spinning can be thus employed. Newly formed fibers, whether formed by melt, wet or dry spinning, are subjected to drawing or stretching. Nylon is cold drawn, whereas the polyesters must be hot drawn. Drawing aligns the molecules, placing them parallel to one another and bringing them closer together. The amount of draw—the draw ratio—determines the decrease in fiber size and the increase in strength, and it varies with intended use.

3 Although each of the synthetic fibers has many individual qualities or characteristics, synthetic fibers as a class possess some common properties. One of the more widely shared characteristics of synthetics is thermoplasticity or sensitivity to heat. Many synthetic fibers are thermoplastic, so that when exposed to heat, they may shrink. To prevent this shrinkage, most synthetic fibers are treated with heat after spinning to "set" them into permanent shape. Not only can fibers be heat—set to make their dimensions permanent but many synthetic fabrics can also be heat-set into pleats, creases, or other permanent shaping.

4 When viewed under the microscope longitudinally, synthetic fibers usually appear as glass rods with either a smooth or striated surface. The fibers appear sufficiently alike that solubility tests are the only means of certain identification. Synthetics tend to be hydrophobic or water resistant, and comfortable when worn next to the skin. On the other hand, many synthetics dry quickly after laundering. Low absorbency also creates difficulties in finishing and dyeing. Static electricity buildup is common among synthetics. Fibers that are more absorbent tend to conduct electricity more readily. More conductive fibers build up electric shocks less readily.

5 Many synthetics are rather smooth and slippery to the touch. Fibers may pill, because their strength prevents the wearing away of the tangled ends. This tendency can be reduced by some of the special texturizing of yarns that is done in manufacturing. It is usually more difficult to remove oil and grease stains from synthetics, as they have an affinity for these substances. Stain removal is

made more difficult by low absorbency. Once the stain has penetrated the fiber, the fiber's resistance to water and other liquids prevents soil removal during laundering or cleaning as the soil is held inside the fiber while the water is kept out.

6 Many of the aforementioned characteristics that present problems to the consumer can be overcome by special finishes or by blending fibers.

7 Two or more different types of textile fibers are often mixed together at the spinning stage. One of the most popular fiber blends is that of cotton and a synthetic fiber called polyester. Shirts, for example, are often one-third cotton and two-thirds polyester, with a growing tendency for cotton's share of the blend to increase. Sheets are often made from cotton and polyester blended in various proportions.

8 Each fiber contributes different properties to the finished blend. The cotton gives absorbency and comfort, as well as surface texture to the fabric. Polyester makes the fabric hardwearing and crease resistant.

9 Another reason for blending fibers together is to balance the cost of each component so that the finished fabric can be sold at a competitive price. The cost of all natural fibers varies with the market demand for them, and also with growing conditions. Bad weather, and attacks by disease carrying insects affect the amount and quality of cotton harvested. If there is less cotton on the market, then the price will go up. Man-made fiber costs can also fluctuate. Most of them are made from oil products, so their market price will be affected by price increases for crude oil.

10 In recent years fabrics of cotton/synthetic fiber blends have been replacing both all-cotton cloths and all-synthetic cloths. Some cotton fiber blends do not use any synthetic fibers. Cotton and wool, and cotton and linen blends are also quite common.

(809 words)

New Words

affinity /əˈfinəti/ *n.* 亲和性
align /əˈlain/ *vt.* 调整，排列整齐
crease /kri:s/ *n.* 折痕
draw /drɔ:/ *vt.* 牵引，牵伸，拉伸
dye /dai/ *vt.* 染色；*n.* 染料
filament /ˈfiləmənt/ *n.* 细丝，灯丝
fluctuate /ˈflʌktʃueit/ *vi.* 波动
hardwearing /ˈhɑ:dweəriŋ/ *adj.* 耐磨的
identification /aidentifiˈkeiʃən/ *n.* 检验，辨认
linen /ˈlinin/ *n.* 亚麻

longitudinally /lɔ:ŋdʒi'tju:dinəli/ adv. 纵向地
microscope /'maikrəskəup/ n. 显微镜
pill /pil/ vi. 做成药丸；服药丸；起球；n. 药丸；弹丸
pleat /pli:t/ n. 褶
polymer /'pɔlimə/ n. 集合物
sensitivity /sensi'tiviti/ n. 灵敏性
share /ʃeə/ n. 份额
slippery /'slipəri/ adj. 光滑的
solubility /sɔlju'biləti/ n. 可溶性
spinneret /'spinərət/ n. 喷丝头
spinning /'spiniŋ/ n. 纺纱
stain /stein/ n. 沾污
striate /'straieit/ adj. 有条纹的
tangle /'tæŋgl/ vt. 缠结
texturize /'tekstʃəraiz/ vt. 变形处理
thermoplastic /θə:məu'plæstik/ adj. 热塑性的
wind-up /waindʌp/ n. 卷绕

Phrases and Expressions

all-cotton	纯棉的，全棉的
all-synthetic	合成的
crude oil	原油
draw ratio	拉伸比
long chain molecule	长链分子
next to the skin	贴身地
melt spinning	熔融纺
public consumption	公众消费

Key Sentences

1. The invention of nylon, and its successful marketing after the war, stimulated the synthesis of additional fibers.

锦纶的发明及其战后在市场上的畅销促进了其他纤维的合成。

2. It was not until the 1920s and 1930s that man first learned how to build long chain molecules from simple substances.

直到20世纪20年代和30年代，人类才开始学会怎样用简单物质制成长链分子。

备注：本句系It引导的强调句，所强调的成分是not until the 1920s and 1930s。that引导的从句中，how to…simple substances形式的不定式短语充当谓语learned的宾语。

3. For those polymers that are adversely affected by heat at or close to their melting temperatures so that they cannot be melt spun, dry spinning or wet spinning can be thus employed.

对于那些在达到或接近其熔化温度时会受到热度的有害影响而不能采用熔融纺丝工艺的聚合物来说，则可采用干法或湿法纺丝工艺。

备注：句中介词短语For those polymers…作状语，其中名词polymers带有一个that引导的定语从句，定语从句本身又带有一个结果状语从句；定语从句中的at or close to their melting temperatures是heat的定语，实际上是两个短语，一个是介词短语at their melting temperature；另一个是形容词短语close to their melting temperature。

Notes

1. 热塑性：物质在加热时能发生流动变形，冷却后可以保持一定形状的性质，在一定温度范围内，能反复加热软化和冷却硬化。

2. 混纺：混纺化纤织物是化学纤维与其他棉毛、丝、麻等天然纤维混合纺纱织成的纺织品。例如，涤棉混纺物是以涤纶为主要成份，采用65%～67%涤纶和33%～35%的棉混纺纱线织成的纺织品，既有涤纶的风格又有棉的特点，在干、湿情况下都有较好的弹性和耐磨性，尺寸稳定，缩水率小，具有挺拔、不易皱折、易洗、快干的特点。

Post-Reading Exercises

• **Reading Comprehension**

Directions: *Work in pairs to decide whether the following statements are true or false according to the passage. Write "T" for True and "F" for False in the space provided.*

_____1) Before the 1920s and 1930s, man already knew how to build long chain molecules from simple substances.

_____2) The diameter of the hole in the spinneret is not a decisive/determining factor of the diameter of the fiber.

_____3) Low absorbency doesn't have any effect on finishing and dyeing.

_____4) There is only one type of textile fibers which are often mixed together at the spinning stage.

_____5) Not all cotton fiber blends use synthetic fibers.

• **Vocabulary**

Direction: *Complete the following sentences with the proper forms of the words given in the box.*

| draw | align | longitudinally | identification |
| dye | tangle | stain | fluctuate |

1) The wounded man's blood _____ the floor red.

2) His idea was to generate a unique _____ tag for each species based on a short stretch of DNA.

3) I _____ my chair up close to the fire.

4) He _____ _____ up the sheets on the bed as she lay tossing and turning.

5) The plant's stem is marked with thin green _____ stripes.

6) If you buy stock in a high tech company that sells things on the Internet, you might expect it to _____ enormously.

7) If you _____ yourself with a particular group, you support them because you have the same political aim.

8) The damp rising from the ground caused the walls to _____ badly.

• **Translation**

Direction: *Translate the following English terms into Chinese.*

1) synthetic fiber_____

2) next to the skin_____

3) all-cotton_____

4) long chain molecule_____

5) draw ratio_____

6) public consumption_____

7) crude oil_____

Unit 3　Textile Yarns and Spinning Technology

PART ONE　Warm-up Activities

Huang Daopo

New Words

Huang Daopo　黄道婆（1245—1330 年），又名黄婆或黄母，宋末元初知名棉纺织家。她向百姓传授先进的纺织技术，推广先进的纺织工具。清代时，黄道婆被尊为布业的始祖。
Li　黎族，中国岭南少数民族之一，以农业为主，妇女精于纺织，"黎锦""黎单"闻名于世。
Wunijing of Songjiang　松江府乌泥泾镇，今上海市华泾镇。

Directions: *Listen to this passage carefully and answer the following questions.*

1. Which dynasty did Huang Daopo live in?

2. Why did Huang Daopo leave her hometown?

3. Where did Huang Daopo go after she left her hometown?

4. From whom did Huang Daopo learn the techniques of planting, spinning, and weaving cotton?

5. List some of the contributions Huang Daopo made to the textile industry in China.

The Story of Silk

New Words

fibroin /ˈfaibrəuin/ *n.* 蚕丝蛋白　　　　　　　　gland /glænd/ *n.* 腺
pry /prai/ *v.* 撬动，撬开

metamorphoses /metə'mɔːfəsiːz/ v. 使变形，使变质
spinneret /'spinəret/ n. （蜘蛛，蚕的）吐丝器　　　sericin /'seərisin/ n. 丝胶蛋白
Louis Pasteur（1822—1895年）路易·巴斯德，法国微生物学家，化学家，近代微生物学的奠基人

Directions: *Listen carefully and decide whether the following statements are true or false. Put "T" for True and "F" for False in the spaces provided.*

_____ 1. It's said that silk was first discovered by the Chinese emperor Huang Di thousands of years ago.

_____ 2. Silkworms are wild caterpillars which cannot be domesticated completely.

_____ 3. Besides mulberry, the silkworm eats eggs in order to spin a cocoon and turn into an adult moth.

_____ 4. The adult moths lay eggs and die in just a few weeks.

_____ 5. The cocoon is made up of silk which is about half a mile long.

_____ 6. The best quality silk is unrolled by hand while the silkworm in the cocoon is kept alive.

_____ 7. In the nineteenth century, French producers got a few cocoons and established a local silk industry.

_____ 8. Pasteur succeeded in saving the silkworms and found that many diseases were caused by bacteria, which marked the beginning of the modern age of medicine.

Intelligent Clothing

Directions: *Listen to this passage carefully and answer the following questions.*

1. What kind of new clothing are scientists developing now?
_____

2. What kind of intelligent clothing has been produced already?
_____

3. What kind of clothing is expected to be produced in the future?
_____

PART TWO   Reading Activities

Passage 1   Yarns and Their Classifications

1 Yarns are long strands of fiber that are used as threads, cords or components in the production of fabrics. Yarns are formed by combining or using textile fibers as a long strand. Yarns may be classified through a variety of means:

(1) by the length of fibers and processes used to form the yarn.

(2) by the twist in the yarn.

(3) by the size of the yarn.

(4) by the number of parts used in the development of the yarn.

(5) by the complexity of the structure of the yarn.

2 Some characteristics will lead to finer yarns while others will lead to bulkier yarns; some produce stronger yarns while others will produce weaker yarns; and some will produce fancier yarns while others will produce more ordinary yarns. By varying characteristics in each of the classifications, yarns take on a unique set of qualities. For some end uses a certain set of properties is highly desirable; yet for other end uses the same set of properties may not be as compelling.

3 Textile products can be enhanced and made more serviceable by selecting yarns that feature a set of desirable characteristics. For example, a designer could create a completely different look by using a fancy, bulky yarn or by using a fine, lightweight, simple yarn. Likewise, the yarns selected can enhance the durability, performance, and overall suitability of products such as uniforms and commercial linens.

4 In this unit, examples of the various classifications of yarns are shown through illustrations and photographs. The examples shown in this unit are fairly simple and easily identifiable. Professional

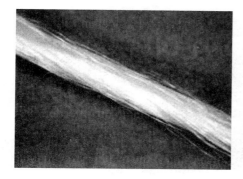

Filament yarn of synthetic fibers

Simple example of nylon being drawn into a fiber

Spun yarn

Silk filament yarn

textile scientists often work with more complex versions of these same basic fibers to create an entire array of unique yarns to meet the aesthetic and performance characteristics needed for modern day commercial, industrial, home and personal uses.

5 Yarns are produced from fibers that are filament or staple. Filament yarns are comprised of a single continuous fiber or group of continuous fibers that are typically twisted to hold them together as a single yarn. With the exception of silk, most filament yarns are derived from manufactured fibers. It should be noted that filament yarns make up only a very small portion of the yarns available in the marketplace. Most yarns are made from staple fibers through a process known as spinning.

6 The spinning process turns a single type of fiber or group of various fibers into a continuous strand by twisting and pulling the fibers into shape. The resultant strand is known as a spun yarn. In preparation for the spinning process, short fibers are carded. Carding directs textile fibers into relatively one direction to create a loosely assembled but somewhat uniform rope known as a "sliver". Slivers of different fibers may be twisted together to form a blended strand in a process called "roving". Strands will continue to be drawn into tighter and narrower strands through the roving process, which continue to redirect the fibers toward a single direction. Ultimately, the strand will be twisted and drawn to its final stage as a spun yarn through the spinning process. In instances where the staple fibers are longer, the fibers may also undergo a process known as "combing". Combing

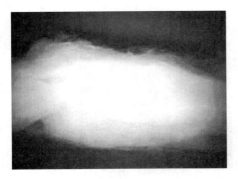

Sliver

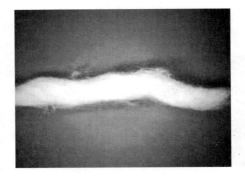

Yarn after roving

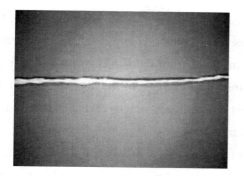
Yarn after combing

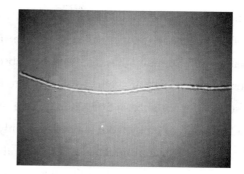
Finished yarn

Commercial spinning equipment

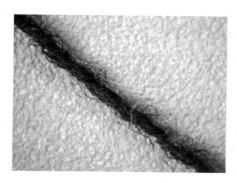
Woolen yarn

Worsted yarn

takes place after carding and is intended to further align the fibers. By further aligning the fibers and removing many of the shorter fibers, the finished yarn becomes smoother, more uniform in shape and stronger.

7 If the spun yarn is comprised of wool fibers or wool blends, specific terms are used to refer to carded or combed yarns. Yarns of longer wool fibers that have been carded are called "woolen yarns", while yarns of shorter wool fibers that have been combed are termed "worsted yarns".

8 Yarn quantities are usually measured by weight in ounces or grams. In the United States, Canada and Europe, balls of yarn for handcrafts are sold by weight. Common sizes include 25g, 50g,

and 100g skeins. Some companies also primarily measure in ounces with common sizes being three-ounce, four-ounce, six-ounce, and eight-ounce skeins. These measurements are taken at a standard temperature and humidity, because yarn can absorb moisture from the air. The actual length of the yarn contained in a ball or skein can vary due to the inherent heaviness of the fiber and the thickness of the strand; for instance, a 50g skein of lace weight mohair may contain several hundred meters, while a 50g skein of bulky wool may contain only 60 meters.

9　There are several thicknesses of yarn, also referred to as weight. This is not to be confused with the measurement of weight listed above. The Craft Yarn Council of America is making an effort to promote a standardized industry system for measuring this, numbering the weights from 1 (finest) to 6 (heaviest). Some of the names for the various weights of yarn from finest to thickest are called lace, fingering, double—knit (or DK), worsted, bulky, and super—bulky. This naming convention is more descriptive than precise; fiber artists disagree about where on the continuum each lies, and the precise relationships between the sizes.

Yarn

10　A more precise measurement of yarn weight, often used by weavers, is wraps per inch (wpi). The yarn is wrapped snugly around a ruler and the number of wraps that fit in an inch are counted.

11　Labels on yarn for handcrafts often include information on gauge, known in the UK as tension, which is a measurement of how many stitches and rows are produced per inch or per centimeter on a specified size of knitting needle or crochet hook. The proposed standardization uses a four-by-four inch/ten-by-ten centimeter knitted or crocheted square, with the resultant number of stitches across and rows high made by the suggested tools on the label to determine the gauge.

12　In Europe textile engineers often use the unit tex, which is the weight in grams of a kilometer of yarn, or decitex, which is a finer measurement corresponding to the weight in grams of 10 kilometers of yarn. Many other units have been used over time by different industries.

(1010 words)

New Words

align /ə'lain/ vt. 使结盟；使成一行；匹配；vi. 排列；排成一行
aesthetic /i:s'θetik/ adj. 美学的
bulky /'bʌlki/ adj. （bulkier 比较级）粗大的，大的
carding /'ka:diŋ/ n. 梳理，梳棉
characteristics /kærəktə'ristiks/ n. 特征
combing /'kəumiŋ/ v. 梳理，精梳

cord /kɔ:d/ n. 绳索；vt. 束缚，用绳子捆绑
Craft Yarn Council of America 美国手工纱线制品委员会
crochet /krəu'ʃei/ hook n. 钩针
feature /'fi:tʃə/ v. 以……为特色；起主要作用
filament /'filəmənt/ n. 长纤维，长丝
fingering /'fiŋgəriŋ/ n. 绒线
gauge /geidʒ/ n. 隔距
handcraft /'hændkrɑ:ft, -kræft/ n. 手工艺，手工
humidity /hju:'midəti/ n. 湿度
mohair /'məuheə/ n. 马海毛
roving /'rəuviŋ/ n. 粗纱
serviceable /'sə:visəbl/ adj. 有用的；可供使用的
skein /'skein/ n. 一束，一个线球
sliver /'slivə/ n. 棉条
spinning /'spiniŋ/ n. 纺纱
staple /'steipl/ n. 短纤纱
stitch /stitʃ/ n. 针脚，线迹；一针；vt. 缝，缝合；vi. 缝，缝合
tension /'tenʃən/ n. 强度
tex /teks/ n. 支数（表示纱线疏密程度的一种单位，每一千米纱线或者纤维所具有的质量克数）
twist /twist/ vt. 捻，捻合
weaver /'wi:və/ n. 织工

Phrases and Expressions

spun yarn   细纱，纺成纱
woolen yarns   粗纺毛线，粗纺毛纱；毛线，绒线
worsted yarns   精纺毛纱

Key Sentences

1. Yarns are long strands of fiber that are used as threads, cords or components in the production of fabrics.

纱线是指在织物生产中被用作线、绳或者生产成分的长束纤维。

2. By varying characteristics in each of the classifications, yarns take on a unique set of qualities.

在每一分类中，纱线通过改变特征可呈现一系列的独特性质。

3. Textile products can be enhanced and made more serviceable by selecting yarns that feature a set of desirable characteristics.

通过选择一系列具有良好特征的纱线，我们可以强化纺织品，使之更加实用。

4. Filament yarns are comprised of a single continuous fiber or group of continuous fibers that are typically twisted to hold them together as a single yarn.

长纤纱线是由单根连续纤维或者一组连续纤维捻合在一起，形成一根纱线。

5. Carding directs textile fibers into relatively one direction to create a loosely assembled but somewhat uniform rope known as a sliver.

梳棉是将织物纤维相对朝某一方向排列，形成一个松散却均一的绳，俗称棉条。

6. Combing takes place after carding and is intended to further align the fibers.

在梳棉之后是精梳，旨在进一步整理纤维。

7. These measurements are taken at a standard temperature and humidity, because yarn can absorb moisture from the air.

因为纱线在空气中会吸湿，所以这些测定都取自于标准温度和湿度。

8. Some of the names for the various weights of yarn from finest to thickest are called lace, fingering, double-knit (or DK), worsted, bulky, and super-bulky.

纱线粗细不同，重量不同，名称就不同。从最细到最粗的纱线被命名为精细、细绒、双面、精纺、粗纺和超粗纺。

9. Labels on yarn for handcrafts often include information on gauge, known in the UK as tension, which is a measurement of how many stitches and rows are produced per inch or per centimeter on a specified size of knitting needle or crochet hook.

手工纱线制品的商标常常包含隔距信息，在英国被熟知为强度，即在一个针织品或者钩针的规定尺寸上每英寸或者每厘米有多少缝线和多少列。

Notes

纱线："纱"和"线"的统称。"纱"是将许多短纤维或长丝排列成近似平行状态，并沿轴向旋转加捻，组成具有一定强力和线密度的细长物体；而"线"是由两根或两根以上的单纱捻合而成的股线。纱线可以用多种天然原料或合成纤维制成，如棉、竹、麻、大豆、羊、骆驼、猫、狗、狐、兔等动物的毛以及各种人造纤维。

Post-Reading Exercises

- **Reading Comprehension**

Directions: *Read the passage and answer the following questions.*

1) According to the passage, what is the definition of "yarns"?

2) How can we classify yarns?

3) How can textile products be made more serviceable?

4) How can a spun yarn be completed?

5) What are sliver, roving and combing?

6) What is the difference between woolen yarns and carded or combed yarns?

7) How can we measure yarn quantities?

8) What is the thickness of yarn?

9) What is a more precise measurement of yarn weight, often used by weavers?

10) What do labels on yarn for handcrafts often include?

- **Vocabulary**

Directions: *Read the following groups of sentences carefully and discuss with your partner how the same word is used with different meanings in each group. Then translate them into Chinese.*

1) align
   a. In return they must demand the right to constrain institutions' behavior and *align* their risks with society's interests.
   b. He neatly *aligned* the flower-pots on the window-sill.
2) card
   a. To achieve this, there is a need to conduct a detailed analysis and *carding*.
   b. I could actually swipe a *card* and generate an electronic receipt via email and then send it out to a person.
3) feature
   a. Now that's not always going to happen, but Nash Equilibrium is kind of a nice *feature*.
   b. This store *features* round-the-clock service.
   c. Can you *feature* wearing a necktie here?

4) fingering

　　a. She stood on the platform, nervously *fingering* her dress.

　　b. He works hard. He practices scales, *fingering*, melody and rhythm.

5) roving

　　a. First, fibers having smooth surfaces require less *roving* twist for satisfactory packaging, handling, and performance in spinning.

　　b. He is a widower with a *roving* eye and the morals of a stray dog.

6) sliver

　　a. If you look at the horizon at sunset, exactly as the last *sliver* of sun disappears, you might see a flash of brilliant green blaze across the sky.

　　b. The log was *slivered* into kindling.

7) staple

　　a. A storage room is set aside especially to deposit fresh bamboo which is the *staple* food for the pandas in their natural habitat.

　　b. That was good for about 10 seconds before it snapped out of that cheap little *staple* they put it in there with.

　　c. Archaeologists have found that the Philistine diet leaned heavily on grass pea lentils, an Aegean *staple*.

8) tension

　　a. In fluids, forces of *tension* and shear can often be ignored.

　　b. Arab moderates may be able to convince the Bush administration that the best way to ease *tension* would be for America itself to be more flexible.

9) twist

　　a. She *twisted* many threads to make this rope.

　　b. The newspaper report *twisted* what the pop singer had said.

10) worst

　　a. *Worst* of all, bonuses are being paid in part from subsidies: this is not a free market, but a perversion of it.

　　b. Color-difference of combing *worsted* spinning is very important to quality control.

- **Translation**

Directions: *Translate the following sentences into English, using the expressions in brackets.*

1）彩色纺纱线在棉针织行业的需求正逐年上升，发展前景看好。（spun yarn）

2）在每一类型中通过改变特征，能够得到具有一系列独特性质的纱线。（take on）

3）我们生产各种比例的粗梳纱线，像山羊绒、羊毛、兔毛、锦纶等，支数为 8～28。（woolen yarn）

_____

4）梳棉之后是精梳，旨在进一步整理纤维。（take place）

_____

Passage 2 Early History and Development of Spinning

1 Although it has yet to be discovered precisely when man first began spinning fibers into yarns, there is much archaeological evidence to show that the skill was well practiced at least 8000 years ago. Certainly, the weaving of spun yarns was developed around 6000 B.C., when Neolithic man began to settle in permanent dwellings and to farm and domesticate animals. Both skills are known to predate pottery, which is traceable to circa 5000 B.C.

2 Man's cultural history goes back about 10,000 to 12,000 years, when some tribes changed from being nomadic forager-hunters, who followed the natural migration of wild herds, to early farmers, domesticating animals and cultivating plants. It is very likely that wool was one of the first fibers to be spun, since archaeologists believe that sheep existed before Homo sapiens evolved. Sheep have been dated back to the early Pleistocene period, around 1 million years ago. The Scotch blackface and the Navajo sheep are present breeds thought to most closely resemble the primitive types. Domesticated sheep and goats date from circa 9000 B.C., grazing the uplands of north Iraq at Zam Chem Shanidar; from circa 7000 B.C., at Jarmo, in the Zagros Mountains of northwest Iran; and in Palestine and south Turkey from the seventh and sixth millennia B.C. Sheep were also kept at Bougras, in Syria, from circa 6000 B.C.

3 We can speculate that early man would have twisted a few fibers from a lock of wool into short lengths of yarn and then tied them together to make longer lengths. We call these staple-spun yarns, because the fibers used are generally referred to as staple fibers. Probably the yarn production would have been done by two people working together, one cleaning and spinning the wool, the other winding the yarn into a ball. As the various textile skills developed, the impetus for spinning continuous knotless lengths would have led to a stick being used, maybe first for winding up the yarn and then to twist and wind up longer lengths, thereby replacing the making of short lengths tied together and needing only one operative. This method of spinning a yarn using a dangling spindle or whorl was widely practiced for processing both animal and plant fibers. Seeds of domesticated flax (Linum usitassimum) and spindle whorls dating back to circa 6000 B.C. were found at Ramad, northern Syria, and also in Samarran villages (Tel-es Swan and Choga Mami) in north Iraq (dated circa 5000 B.C.). In Egypt, at Neolithic Kom, in Fayum, stone and pottery whorls of about 6000 B.C. have been discovered, while at the predynastic sites of Omari, near Cairo, and Abydos, both circa 5500 B.C., flax seeds, whorls, bone needles, cloth, and matting have been found.

4 Flax was probably the most common ancient plant fiber made into yarns, though hemp was also used. Although flax thread is mentioned in the Biblical records of Genesis and Exodus, its antiquity is even more ancient than the Bible. A burial couch found at Gorigion in ancient Phrygia and dated to be late eighth century B.C., contained twenty layers of linen and wool cloth, and fragments of hemp and mohair. Cotton, native to India, was utilized about 5000 years ago. Remnants

of cotton fabric and string dating back to 3000 B. C. were found at archaeological sites in Indus in Sind (India). Many of these fibers were spun into yarns much finer than today's modern machinery can produce. Egyptian mummy cloth was discovered that had 540 threads per inch in the width of the cloth. Fine-spun yarns, plied threads, and plain-weave tabby cloths and dyed garments, some showing darns, were also found in the Neolithic village of Catal Huyuk in southern Turkey.

5 The simple spindle continued as the only method of making yarns until around A .D. 1300, when the first spinning wheel was invented and was developed in Europe into "the great wheel" or "one-thread wheel." The actual mechanization of spinning took place over the period 1738 to 1825 to meet the major rise in the demand for spun yarn resulting from then spectacular increase in weaving production rates with the invention of the flying shuttle (John Kay, 1733). Pairs of rollers were introduced to thin the fiber mass into a ribbon for twisting (Lewis Paul, 1738); spindles were grouped together to be operated by a single power source—the "water frame" (Richard Arkwright, 1769), the "spinning jenny" (James Hargreaves, 1764—1770) and the "mule" (Samuel Crompton) followed by the "self-acting mule" by Roberts (1825). In 1830, a new method of inserting twist, known as cap spinning, was invented in the U.S. by Danforth. In the early 1960s, this was superseded by the ring and traveler, or ring spinning, which, despite subsequent inventions, has remained the main commercial method, and is now an almost fully automated process.

Spinning wheel

6 Today, yarn production is a highly advanced technology that facilitates the engineering of different yarn structures having specific properties for particular applications. End uses include not only garments for everyday use and household textiles and carpets but also sports clothing and fabrics for automotive interiors, aerospace, and medical and healthcare applications. A detailed understanding of how fiber properties and machine variables are employed to obtain yarn structures of appropriate properties is, therefore, an important objective in the study of spinning technology.

(895 words)

New Words

Abydos /ə'baidɔs/ n. 阿比多斯（埃及古城），阿拜多斯（小亚细亚古城）
be superseded by 为……取代
Bougras /'bəugræs/ n. 布格拉斯
circa /'sə:kə/ prep. 大约于；adv. 大约
darn /da:n/ v. 织补，缝补；n. 织补
Exodus /'eksədəs/ n. 《出埃及记》（《圣经》第二卷）
Fayum /'faijum/ n. 法雍（埃及一地区）
forager /'fɔridʒə/ n. 搜寻（食物）的人，觅食者
Genesis /'dʒenisis/ n. 《创世纪》（基督教《圣经》的首卷）[略作 Gen.]
Homo sapiens /'həuməu'seipiənz; 'sæpienz/ n. 智人（现代人的学名）；人类
impetus /'impitəs/ n. 动力；促进；冲力
Indus /'indəs/ n. 印度河
knotless /'nɔtlis/ adj. 无结的
matting /'mætiŋ/ n. 席子；编席的原料；
　　　　　　　　v. 纠缠在一起；铺席于……上；使……无光泽
Navajo /'nævəhəu/ n. 纳瓦霍人（美国最大的印第安部落）
Neolithic /ˌni:əu'liθik/ adj. [古]新石器时代的；早先的
operative /'ɔpərətiv/ adj. 有效的；运转着的；从事生产劳动的；n. 侦探；技工
Palestine /'pælistai/ n. 巴勒斯坦
Phrygia /'fridʒiə/ n. 佛里吉亚
Pleistocene period n. [地质]更新世时期
Scotch blackface n. 苏格兰黑面羊
spindle /'spindl/ n. 纺锤，锭子；细长的人或物；adj. 锭子的，锭子似的；细长的
spinning /'spiniŋ/ n. 纺纱
Syria /'si:riə/ n. 叙利亚共和国
tabby /'tæbi/ n. 平纹；斑猫；长舌妇；adj. 起波纹的；有斑纹的；vt. 使起波纹
weaving /'wi:viŋ/ n. 织造
whorl /wə:l/ n. 螺纹；轮生体；涡；vt. 使成涡旋；vi. 盘旋
Zagros Mountains n. 扎格罗斯山脉

Phrases and Expressions

household textile　　　　　　家用纺织品
mummy cloth　　　　　　　木乃伊裹布

ring spinning	环锭纺纱
spinning jenny	多轴纺织机
spinning technology	纺纱工艺
staple fibers	纺纱用的人造短纤维
staple-spun yarns	短纤纱

Key Sentences

1. We can speculate that early man would have twisted a few fibers from a lock of wool into short lengths of yarn and then tied them together to make longer lengths. We call these staple-spun yarns, because the fibers used are generally referred to as staple fibers.

我们可以推测，早期的人们从一簇羊毛中捻取一些纤维形成长度很短的纱线，然后把它们打结形成较长的纱线。我们称这些纱线为短纤纱，因为所用的纤维通常被称为短纤维。

2. As the various textile skills developed, the impetus for spinning continuous knotless lengths would have led to a stick being used, maybe first for winding up the yarn and then to twist and wind up longer lengths, thereby replacing the making of short lengths tied together and needing only one operative.

各种纺织技术的发展促进连续、无结纱线的纺制，这导致辊的采用，可能最初用于卷绕纱线，而后用于加捻并卷绕长度更长的纱线，不必再将短纱结在一起，而且一次性操作完成。

3. Many of these fibers were spun into yarns much finer than today's modern machinery can produce.

这些纤维中很多种被纺成纱线，比现代机器能纺成的纱线细得多。

Post-Reading Exercises

- **Reading Comprehension**

Directions: *Read the passage and complete the following sentences.*

1) Archaeologists believe that _____ existed before _____ _____ evolved. _____ have been dated back to the early _____ _____, around 1 million years ago.

2) Probably the yarn production would have been done by two people working together, one _____ _____ _____ the wool, the other _____ the yarn into a ball.

3) _____ was probably the most common ancient plant fiber made into yarns, though _____ was also used. Although _____ _____ is mentioned in the Biblical records of Genesis and Exodus, its _____ is even more ancient than the Bible.

4) In the early 1960s, this _____ _____ _____ the ring and traveler, or _____ _____, which, despite other subsequent later inventions, has remained the main _____

method and is now an almost fully _____ process.

5) Today, _____ _____ is a highly advanced technology that _____ the engineering of different yarn structures having _____ _____ for particular applications.

- **Vocabulary**

Directions: *Complete the following sentences with the proper forms of the words and phrases given in the box.*

| domesticate | facilitated | permanent | primitive |
| resemble | speculate | staple | variable |

1) When people _____ wild animals or plants, they brought them under control and used them to produce food or as pets.

2) At the end of the probationary period this intern will become a (an) _____ employee.

3) Generally, rice is the _____ food of more than half the world's population.

4) The press acknowledged that the economic recovery had been _____ by the Prime Minister's tough stance.

5) Some of the commercially produced leather _____ genuine one not only in exterior but also in touch.

6) There was a bit of a wind and it was blowing onshore, _____, but quite strong.

7) It is a (an) _____ instinct of all animals to flee a place of danger.

8) From the outside syndrome, the doctor _____ that he died of cardio-vascular disease.

- **Translation**

Directions: *Translate the following Chinese terms into English.*

1）纺纱工艺 _____
2）环锭纺纱 _____
3）短纤纱 _____
4）家用纺织品 _____
5）多轴纺织机 _____
6）木乃伊裹布 _____
7）无结纱线 _____
8）纺纱用的人造短纤维 _____

Passage 3 Fiber Spinning

1 Spinning is the twisting together of fibers to form yarn (or thread, rope, cable). Earlier fiber was spun by hand using simple tools like spindle and distaff. Later the use of spinning wheel gained importance. Industrial spinning started in the 18th century with the beginning of the Industrial Revolution. Hand-spinning remains a popular handicraft.

Industrial spinning

2 Fibers cannot be used to make clothes in their raw form. For this purpose, they must be converted into yarns. The process used for yarn formation is spinning. Spinning by hand was a slow and laborious process. Thus, many implements and methods were invented for making it faster and simpler. Eventually, the techniques were refined and industrial spinning started manufacturing yarn in various ways. The methods selected depend upon the factors such as the manufacturer's preference of equipment, the economic implications, the fibers to be used and the desired properties of yarn to be produced. Ring method is the oldest and the most used technique. Open-end spinning is another important method. The basic manufacturing process of spinning includes carding, combing, drafting, twisting and winding. As the fibers pass through these processes, they are successively formed into lap, sliver, roving and finally yarn. A brief description of the journey from fibers to yarns will help in understanding industrial spinning in a better way.

3 The raw fiber arrives at a spinning mill as a compressed mass which goes through the processes of blending, opening and cleaning. Blending is done to obtain uniformity of fiber quality. Opening is done to loosen the hard lumps of fiber and disentangle them. Cleaning is required to remove the trash such as dirt, leaves, burrs and any remaining seeds. Carding is the initial straightening process which puts the fiber into a parallel lengthwise alignment. This makes the tangled mass of fiber ready to produce yarn. Now the fiber is called "Lap". The lap is treated for removing the remaining trash, disentangling and molding it into a round rope—like mass called a "Sliver". The sliver is then straightened again, a process which is called "Combing". In it, fine-toothed combs continue straightening the fibers until they are arranged in such a parallel manner that the short fibers are completely separated from the longer fibers. This procedure is not required for man-made staple fiber because they are cut into predetermined uniform lengths. This process forms a "comb sliver" made of the longest fibers. The combing process is identified with better quality because long staple yarn produces stronger, smoother and more serviceable fabrics.

4 Drawing pulls the staple lengthwise over each other. As a result, longer and thinner slivers are produced. After several stages of drawing

Ring spinning

out, the sliver is passed to the spindles where it is given its first twist and is then wound on bobbins. "Roving" is the final product of the several drawing-out operations. It is the preparatory stage for the final insertion of twist. Till now, enough twist is given for holding the fibers together but it has no tensile strength. It can break apart easily with a slight pull. The roving, on bobbins, is placed in the spinning frame, where it passes through several sets of rollers running at high speed. Finally the "Yarn" is produced of the sizes desired.

Spinning Machines Traditional VS. Modern Techniques

5 Hand spinning was replaced by powered spinning machines which were very fast. Initially machines were powered by water or steam and then by electricity. The spinningenny, a multi-spool spinning wheel significantly reduced the amount of work required to produce yarn. A single worker was now able to work eight or more spools at a time.

6 Then came the spinning frame which produced a stronger thread than the spinning jenny. As it was too large to be operated by hand, a spinning frame powered by a waterwheel was invented. It was then called the water frame. The elements of the spinning jenny and water frame were combined to create the spinning mule.

(664 words)

New Words

bobbin /'bɔbin/ *n.* 纱筒
burrs /bə:rs/ *n.* 毛边；过火砖（burr 的复数形式）；
　　　　　　v. 从……除去毛刺；在……上形成毛边
distaff /'dista:f, -tæf/ *n.* 卷线杆
drawing /'drɔ:iŋ/ *n.* 牵伸
lap /læp/ *n.* 一圈；*vt.* 使重叠；拍打；包围；*vi.* 重叠；轻拍；围住
lengthwise /'leŋθwaiz/ *adj.* 纵长的；*adv.* 纵长地
spool /spu:l/ *vt.* 缠绕；卷在线轴上；*n.* 线轴；缠线框
valve spool 滑阀；阀槽
spool valve *n.* 短管阀
enwind *vt.* 缠绕；卷在线轴上

Phrases and Expressions

break apart　　　　　　　　　　使……分裂开
fine-toothed comb　　　　　　　细齿梳子
open-end spinning　　　　　　　自由端纺纱

ring method	环锭纺
spinning frame	细纱机，精纺机，精纱机
spinning mill	纺纱厂
spinning mule	纺纱机
tensile strength	抗张强度
water frame	水力纺纱机

Key Sentences

1. Spinning is the twisting together of fibers to form yarn (or thread, rope, or cable).
纺纱是将纤维缠绕在一起而形成纱线（或线、绳或索）。

2. The basic manufacturing process of spinning includes carding, combing, drafting, twisting and winding.
纺纱的基本加工过程包括粗梳、精梳、前伸、加捻和卷绕。

3. As the fibers pass through these processes, they are successively formed into lap, sliver, roving and finally yarn.
纤维经过这些加工后，分别形成棉卷、棉条、粗纱及最终的纱线。

Notes

纺纱（spinning）：纺纱过程一般包括原料准备、开松、梳理、除杂、混和、牵伸、并合、加捻以及卷绕等作用。

Post-Reading Exercises

• **Reading Comprehension**

Directions: *Work in pairs to decide whether the following statements are true or false according to the passage. Write "T" for True and "F" for False in the space provided.*

1) _____ Industrial spinning started in the 19th century with the beginning of the Industrial Revolution.

2) _____ Open-end spinning is the oldest and the most used technique. Ring method is another important technique.

3) _____ Cleaning is required to remove the trash such as dirt, leaves, burrs and any remaining seeds. Carding is the initial straightening process which puts the fiber into a parallel lengthwise alignment.

4) _____ "Combing" is the final product of the several drawing—out operations. It is the

preparatory stage for the final insertion of twist.

5) _____ The spinning jenny, a multi-spool spinning wheel significantly reduced the amount of work required to produce yarn.

- **Vocabulary**

Direction: *Complete the following sentences with the proper forms of the words given in the box.*

| alignment | bobbin | burrs | initially |
| lap | lengthwise | disentangle | successively |

1) Nor is it easy to _____ the effects of climate change from those of avoidable failures in policy.

2) The thread has not been passed through the notch of the _____ case tension spring.

3) Because of his good public image he has _____ gained office.

4) All surfaces are free of _____ to prevent the accumulation of dirt.

5) If something is not in _____ with your life goal, drop it or set it aside.

6) Morrison should try to get away from them in the third _____.

7) Even if the genes themselves are not _____ hazardous, you do not know how they are going to evolve.

8) Look for the center of your fabric by folding it in half sideways and then again _____.

- **Translation**

Directions: *Translate the following Chinese terms into English.*

1）手纺车 _____

2）纺织厂 _____

3）开端纺纱、自由端纺纱 _____

4）平衡的纵向对齐 _____

5）矫直过程 _____

6）拔出操作 _____

Unit 4　Woven Fabrics and Weaving Technology

PART ONE　Warm-up Activities

Chinese Archaeologists Make Ground-breaking Textile Discovery

New Words

archaeologist /ˌaːkiˈɔlədʒist/ n. 考古学家	vermilion /vəˈmiliən/ n. 朱砂
handicraft /ˈhændikrɑːft/ n. 手工艺品	preservation /ˌprezəˈveiʃn/ n. 保存，储藏
excavation /ˌekskəˈveiʃn/ n. 挖掘	aristocrat /ˈæristəkræt/ n. 贵族
ground-breaking adj. 开创性的	unearth /ʌnˈəːθ/ vt. 出土
tissue /ˈtiʃuː/ n. ［生］组织	infrared /ˌinfrəˈred/ n. 红外线

Directions: *Listen to the recording and choose the most appropriate answer to each of the following questions.*

1. The largest piece of fabric found in the tomb is _____ long.
   A. 130 m　　　B. 130 cm　　　C. 52 m　　　D. 52 cm
2. Historical records show the Europeans learned how to produce vermilion in _____.
   A. the 8th Century　　B. the 18th Century
   C. the 7th Century　　D. the 17th Century
3. The tomb where the fabrics were found is believed to date back to the _____ Dynasty.
   A. Tang　　B. Western Zhou　　C. Eastern Zhou　　D. Shang
4. Which of the following options was not unearthed from the tomb?
   A. Copperware.　　B. Jade.
   C. Gold.　　D. Handicrafts made from silver.
5. Experts say the discovery is unique because the skeletons were well preserved in an area where the soil was acidic and _____.
   A. where many other valuable archeological objects had been found before
   B. where no valuable archeological finding has ever been found before
   C. where the soil was unsuited to the preservation of human bodies
   D. where the soil was suitable for the preservation of human bodies

History of Clothing and Textiles

New Words

spun /spʌn/ *adj.* 纺成的，拉成丝状的	loop /luːp/ *vt.* 使（绳等）成圈
Stone Age 石器时代	mummification /ˌmʌmifiˈkeiʃn/ *n.* 木乃伊化
Neolithic /ˌniːəˈliθik/ period 新石器时代	

Directions: Listen to "History of Clothing and Textiles" twice and fill in each blank with the information you get from the recording.

The wearing of clothing is exclusively a human 1. _____ and is a feature of most human societies. It is not known when humans began wearing clothes. 2. _____ believe that animal skins and vegetation were adapted into coverings as 3. _____ from cold, heat and rain, especially as humans migrated to new climates. Alternatively, covering may have been 4. _____ first for other purposes, such as magic, decoration, cult, or prestige, and later found to be 5. _____ as well.

Clothing and 6. _____ have been important in human history and reflects the materials available to a 7. _____ as well as the technologies that it has mastered. The social 8. _____ of the finished product reflects their culture.

Textiles, defined as felt or spun 9. _____ made into yarn and subsequently netted, looped, knit or woven to make fabrics, appeared in the Middle East during the late stone age. From 10. _____ times to the present day, methods of textile production have continually 11. _____, and the choices of textiles 12. _____ have influenced how people carried their possessions, clothed themselves, and 13. _____ their surroundings.

Evidence exists for production of linen cloth in Ancient Egypt in the Neolithic period, c. 5500 B.C. 14. _____ of domesticated wild flax, probably an 15. _____ from the Levant, is documented as early as c. 6000 B.C. Other bast fibers including rush, reed, palm, and papyrus were used alone or with linen to make rope and other textiles. Evidence for 16. _____ production in Egypt is scanty at this period.

17. _____ techniques included the drop spindle, hand-to-hand spinning, and rolling on the thigh. Yarn was also spliced. A horizontal ground 18. _____ was used prior to the New Kingdom, when a vertical two-beam loom was introduced, probably from Asia.

Linen bandages were used in the burial 19. _____ of mummification, and art depicts Egyptian men wearing linen kilts and women in narrow dresses with 20. _____ forms of shirts and jackets, often of sheer pleated fabric.

What Do You Wear Every Day?

Directions: *Listen to this passage and match Column A with Column B.*

Column A	Column B
Sunday	a Confidence is key.
Monday	b Fitting in is way overrated.
Tuesday	c Gold sequins go with everything.
Wednesday	d A shiny tiger.
Thursday	e Developing your own unique personal style.
Friday	f Embrace your inner child.
Saturday	g Color is powerful.

PART TWO Reading Activities

Passage 1 Recent Developments: Weaving Technology

1 Fabric manufacturers are under pressure more than ever. Weavers around the world are fighting against higher costs in terms of labor, raw material, production time and reduce energy consumption to remain competitive. In turn, the machinery suppliers are challenged to provide up-to-date machinery.

2 In times of increasing energy costs, it is of utmost interest for fabric producers to use weaving machines that offer reduced energy consumption.

Adaptive Relay Valve Drive

3 Picanol NV, Belgium, provided the following information about its Adaptive Relay Valve Drive, which is featured on its OMNI plus 800 air-jet weaving machine.

4 Adaptive Relay Valve Drive (ARVD) automatically adapts the closing timing of the relay nozzle valves to the behavior of the filling yarn. Instead of applying one setting for all the different picks, ARVD applies the best relay valve timing for each individual pick, thus reducing the overall air consumption.

The Adaptive Relay Valve Drive (ARVD) for Picanol's OMNIplus 800 air-jet weaving machine lowers air consumption and thus reduces energy costs.

5 At each insertion cycle, the winding timings from the prewinder are compared with a reference value by the machine's microprocessor. Since the filling insertion speed varies from one pick to another, it is not necessary for the relay nozzles to blow for the same amount of time at each pick. Consequently, the machine adapts the closing timings for the relay nozzle valves automatically. For a fast pick, the relay nozzle valves do not stay open very long. For a slow pick, they stay open longer... This is done from the second relay nozzle valve onwards.

6 Picanol notes that decreased air consumption reduces energy costs, and reduced blowing on the filling yarn reduces the number of broken picks. Also, Pick Repair Automation has a higher success rate because the type of filling stops changes.

Air-Jet Weaving

7 Italy-based Itema Weaving has recently upgraded its Sulzer Textile ™ L5500 air-jet weaving

machine, suited primarily for applications such as quality apparel and home textile fabrics made with natural or man-made fibers or blends.

8   Key benefits of this machine are said to be the fabric quality and low running costs. According to Itema Weaving, the L5500's strength is its competitiveness in terms of its capacity to conveniently weave fabrics that comply with superior quality standards, while also maintaining a high degree of efficiency even at top performance levels. The company adds that "conveniently" also means producing with reduced off—quality rates and reduced air consumption per meter of fabric, which improves profits. The L5500's RTC (Real Time Controller) function enables the machine to adapt to various weaving conditions, thereby obtaining significant air-consumption savings.

9   Targeted customers are weavers interested in superior production to enable entrance into new, highly profitable markets. The L5500 can be adapted to different production requirements. Easy operation, widest range of application and shortest setting times are key factors.

10   Recent upgrades in the L5500 technology have increased the machine's efficiency and performance. For standard weaving, the maximum speed now is 1,200 revolutions per minute, and the maximum weaving width is 4,000 millimeters (mm). This is equal to a production capacity in excess of 2,500 meter per minute weft insertion rate.

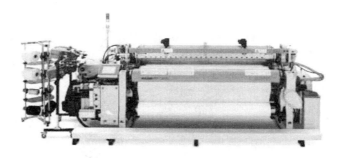

Itema Weaving's Sulzer Textil ™ L5500 air-jet machine features several recent upgrades.

Drawing–in: An Important Cost Factor

11   Switzerland-based Stäubli AG reports that automatic drawing-in machines for the weaving harness have long helped weaving mills around the world stay competitive. Automatically drawn-in warps are characterized by zero defect and high quality, and they are available as required for production in a fraction of the time compared with manual drawing-in.

12   According to Stäubli, the upgraded Safir automatic drawing-in machine offers new opportunities regarding flexibility, thanks to the refinement of proven system components from the Delta line combined with established state-of-the-art technologies, particularly the opal leasing machine. Since its introduction at ITMA 2007 in Munich, Germany, and after several upgrades, the machine can be configured to draw in one or two warp beams, each having up to eight thread layers.

A camera system checks the yarn to be drawn in during each cycle and ensures against drawing-in of double threads or threads of the wrong color.

13 Stäubli reports there also is flexibility with regard to the weaving harnesses that can be used, as virtually all heddles used in shaft weaving can be handled freely. Heddle distribution can be programmed on up to 28 frames. It also is possible to use two different types of drop wires for the same weaving harness—an application that is especially interesting for terry weaving with upper beams and varying drop wire weights for the ground threads and pile threads. With virtually unrestricted application potential, Stäubli says Safir sets new standards in automatic drawing-in. The machine also features user-friendly ergonomics and convenient operation using a color touchscreen.

14 Warp-tying is another cost factor in the weaving mill. Stäubli reports its Magma warp-tying machine is especially suitable for tying coarse yarns. A patent-pending system that works without yarn—specific settings enables separation of threads to be tied at the lease, which considerably simplifies operation and changing from one application to the next. A built-in camera system monitors the separation of the threads, thus eliminating doubled threads. Magma also can be set easily to tie double knots, and therefore can tie even very slick yarns reliably, according to the company.

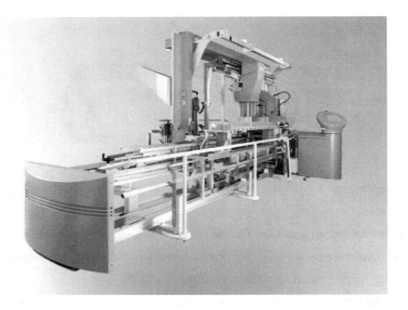

Stäubli's upgraded Safir automatic drawing-in machine can be configured to draw in one or two warp beams, each having up to eight thread layers.

15 The Magma warp-tying machine for coarse yarns complements Stäubli's Topmatic tying machine line. The machines feature a new, patented separating system that reliably separates threads from the lease completely without using thread—specific separating elements and without making special adjustments. Optical sensors check every separated yarn pair before tying, thus preventing

a false double yarn from being tied. The fault can be corrected simply. Depending on yarn material, single or double knots can be tied, as selected by a simple push of a button. Optimization of the tying rate is accomplished using the adjustment wheel.

16 Magma can tie a range of yarns including wool, cotton, linen and other staple yarns, as well as mono- and multifilaments, polypropylene ribbons, and many other yarn types. The yarn count for staple fibers ranges from Ne 0.3 to 50 for warps with 1∶1 lease. The machine comes equipped with an optical double-yarn sensor, and single or double knots can be selected easily. The length of knot ends is variable, with a minimum of 5 mm. The tying frame can be used on all Stäubli type TPF3 tying frames, and easy maintenance is guaranteed, with only regular lubrication needed.

(1091words)

New Words

configure /kən'fiɡə/ v. 使成形
drawing-in n. 穿经
eliminate /i'limineit/ v. 消除，清除；排除（……的可能性）；不考虑，淘汰；杀害，消灭
feature /'fi:tʃə/ n. 特征，特点；容貌，面貌；（期刊的）特辑；故事片；
 vt. 使有特色；描写……的特征；vi. 起主要作用；作重要角色
filling /'filiŋ/ n. （织品的）纬纱
fraction /'frækʃn/ n. 小部分，碎片，片断
harness /'ha:nis/ n. ［印，纺］综绕；（提花机上的）通丝
heddle /hedl/ n. 纺织机的综线
manual /'mænjuəl/ adj. 手的；手制的，手工的；n. 手册；指南
manufacturer /ˌmænju'fæktʃərə/ n. 制造商
microprocessor /ˌmaikrəu'prəusesə/ n. 微处理器
millimeter /'miliˌmi:tə/ n. 毫米
prewinder /pri:'waində/ n. 预络纱机
pick /pik/ n. 纬纱；投梭
refinement /ri'fainmənt/ n. 精炼，提纯，净化；改良品；改良，极致；优雅高贵的动作
revolution /ˌrevə'lu:ʃn/ n. 革命，剧烈的变革；回转，绕转，旋转，转数；周期；一转
shaft /ʃa:ft/ n. 柄，轴；矛，箭；vt. 给……装上杆柄；
slick /slik/ adj. 光滑的，滑溜的；熟练的，灵巧的
state-of-the-art adj. 最新型的，最先进的，顶尖水准的，使用了最先进技术的
up-to-date adj. 最新（式）的；现代化的，尖端的；直到最近的
warp /wɔ:p/ n. ［纺］经（纱）

weft /weft/ n. （纺织物的）纬线；织品；薄云层
winding /'waindiŋ/ n. 络纱

Phrases and Expressions

automatic drawing-in machine	自动穿经机
drop wire	停经片
filling yarn	纬纱
ground threads	地线
heddle distribution	综片分布
man-made fiber	人造纤维
off-quality rate	不达标产品概率
opal leasing machine	分经机
pile threads	绒线
quality apparel	优质服装
relay nozzle valve	辅喷阀
shaft weaving	轴织
targeted customers	目标客户
terry weaving	毛圈织造
tying coarse yarns	接粗纱
warp beam	经轴
weaving harness	织造综

Key Sentences

1. Picanol NV, Belgium, provided the following information about its Adaptive Relay Valve Drive, which is featured on its OMNI plus 800 air-jet weaving machine.

比利时必佳乐公司推出了自适应辅喷阀驱动系统，用于 OMNI plus 800 喷气织机。

2. Adaptive Relay Valve Drive (ARVD) automatically adapts the closing timing of the relay nozzle valves to the behavior of the filling yarn.

自适应辅喷阀驱动系统能根据纬线的运动自动适应辅喷阀的闭合时间。

3. Instead of applying one setting for all the different picks, ARVD applies the best relay valve timing for each individual pick, thus reducing the overall air consumption.

自适应辅喷阀驱动系统不是只用一种设置应对不同的投梭，而是针对每个独立的纬纱应用最佳的辅喷阀时间，由此降低了整体的空气消耗。

4. Since the filling insertion speed varies from one pick to another, it is not necessary for the

relay nozzles to blow for the same amount of time at each pick.

由于纬纱嵌入速度根据纬数的不同而变化，辅喷阀无需在每次投梭时都花同样多的鼓风时间。

5. Italy-based Itema Weaving has recently upgraded its Sulzer Textil ™ L5500 air-jet weaving machine, suited primarily for applications such as quality apparel and home textile fabrics made with natural or man-made fibers or blends.

意大利意达织造公司最近升级了 Sulzer Textil ™ L5500 喷气织机，主要适用的领域包括使用天然纤维或化学纤维或混纺纤维生产优质服装和家居纺织品。

6. According to Itema Weaving, the L5500's strength is its competitiveness in terms of its capacity to conveniently weave fabrics that comply with superior quality standards, while also maintaining a high degree of efficiency even at top performance levels.

据意达织造公司称，L5500 的优点是具有竞争优势，它能够便捷地织造出符合最高质量标准的面料，甚至在最佳表现水平时还能够维持高效率运行。

7. The L5500's RTC (Real Time Controller) function enables the machine to adapt to various weaving conditions, thereby obtaining significant air-consumption savings.

L5500 的实时控制器功能让它可以适应不同的织造条件，由此极大地节省空气消耗。

8. Easy operation, widest range of application and shortest setting times are key factors.

操作简单、应用广泛和设置时间短，这都是它的关键指标。

9. Recent upgrades in the L5500 technology have increased the machines' efficiency and performance.

L5500 技术上的最新升级提高了该机器的效率和性能。

10. Automatically drawn-in warps are characterized by zero defect and high quality, and they are available as required for production in a fraction of the time compared with manual drawing-in.

与手动穿经相比，自动穿经的经纱的特点是零缺陷和质量上乘，当要求在一小部分时间内生产，它们也是可行的。

11. According to Stäubli, the upgraded Safir automatic drawing-in machine offers new opportunities regarding flexibility, thanks to the refinement of proven system components from the Delta line combined with established state-of-the-art technologies, particularly the Opal leasing machine.

据史陶比尔公司称，升级后的 Safir 自动穿经机在灵活性上带来更多机会，这要感谢 Delta 产品线上的验证系统部件的改进以及成熟的先进技术，特别是 Opal 分经机。

12. The machines feature a new, patented separating system that reliably separates threads from the lease completely without using thread-specific separating elements and without making special adjustments.

该机器的特点是带有获得专利的新分离系统，能够可靠地从分经处分离纱线，完全不使用纱线特殊分离部件，也无需做特殊调整。

Notes

1. 比利时必佳乐公司（Picanol NV，Belgium）：比利时必佳乐公司是当今世界最具影响力的织机厂商之一。自1936年成立以来，必佳乐公司一直致力于将最先进的织机技术带给全世界，在喷气和剑杆织机技术方面处于全球领先地位。迄今为止，已经有27万多台必佳乐的织机运行于世界110多个国家的3000多家纺织企业。

2. 意大利意达织造集团（Italy-based Itema Weaving）：意大利意达织造集团源于1834年在瑞士温特图尔（Winterther）成立的苏尔寿公司，目前意大利意达织造集团在全球安装运行的机台数超过200,000台，在全球超过100个国家的代表机构提供服务，拥有著名织机品牌，如苏泰丝、舒美特和范美特，号称是除剑杆织机和喷气织机外唯一能提供片梭织机的厂家，唯一能提供三种引纬方式织机的供应商。

3. 瑞士史陶比尔集团（Switzerland-based Stäubli）：史陶比尔集团的总部位于瑞士的飞飞空（Pfaffikon），创立于1892年，是纺织机械、工业连接器和工业机器人这三大领域机电一体化解决方案的专业供应商。一个多世纪以来，史陶比尔集团以其优质的解决方案和操作程序闻名于世。

4. 喷气织机（air-jet loom）：用喷射出的压缩气流对纬纱进行牵引，将纬纱带过梭口。喷气织机最大特点是车速快、劳动生产率高，适用于平纹和纹路织物、细特高密织物和批量大的织物的生产。喷气织机属于无梭织机（shuttleless loom），本单元Passage 2将具体介绍无梭织机的四个主要类别。

Post-Reading Exercises

- **Reading Comprehension**

Directions: *Read the passage and answer the following questions.*

1) According to the passage, why are fabric manufactures under pressure now more than ever?

2) With energy cost increasing, what is of utmost interest for fabric producers?

3) How does ARVD reduce the overall air consumption?

4) What applications are Itema's updated air-jet weaving machine so primarily suited for?

5) What are the key benefits of the updated air-jet weaving machine?

6) Who are the targeted customers of the L5500?

7) What are the maximum speed and weaving width of updated L5500 for standard weaving?

_____

8) What main characteristics do automatically drawn-in warps have?

_____

9) Which machine does Stäubli recommend for tying coarse yarns?

_____

10) Make a list of the yarns Magma can tie.

_____

- **Vocabulary**

Directions: *Fill in each of the following blanks with the appropriate form of the word in brackets.*

1) The rifle is able to fire continuously because the bullets are loaded _____. (automatic)

2) Chinese consumers seem to have even more of a taste for _____ than most. (vary)

3) It had never crossed my mind that there might be a serious _____. (consequent)

4) With almost half of household wealth tied up in property, will falling home prices make China _____ even less? (consumption)

5) Credit cards are a great _____, but it can be all too easy to let your debt slide out of control. (convenient)

6) The organization defines active blogs as those which are _____ at least once a month. (update)

7) In using to use public policies to reduce inequality, the goal should be to achieve equality without reducing the _____ of the economy. (efficient)

8) His face had something of the youthful, optimistic look peculiar to the _____ English type. (refinement)

9) _____ all of her savings go to pay for tuition at an expensive school for her fourth-grade daughter. (virtual)

10) "In the long term, economic restructuring requires market competition and the _____ of weaker players", he said. (eliminate)

- **Translation**

Directions: *Translate the following sentences into English, using the expressions in brackets.*

1）操作简便，应用广泛和设置时间短是该机器的关键指标。（wide range of）

_____

2）专家预测多项纺织品和服装标准即将更新。（update）

3）这种喷气织机的突出特点是能够生产高品质面料，同时运行成本低廉。（be characterized by）

4）生产商致力于应用最新织造机器降低生产成本。（state-of-the-art）

5）无梭织机种类众多，如剑杆织机、片梭织机、喷气织机和喷水织机等，人们可根据实际需要进行选择。（range from...to）

Passage 2   Shuttleless Looms

1   For centuries, the basic loom operated with a shuttle to lay the filling, or picks. The speed with which the shuttle is sent back and forth is limited by the mass of shuttle—usually about 200 picks per minute. Manufactures have long sought a way to replace the shuttle and increase the speed of weaving.

2   Shuttleless looms were developed as a way to weave without the shuttle. Because they both increased productivity and lowered noise levels, they were widely adopted. Shuttleless weaving machines moved 17 percent more fabric in 1987 than they did in 1982, and Textile World predicts that shuttle looms will be outnumbered by shuttleless weaving machines by the early 1990s. The major types of shuttleless looms are water jet, air jet, rapier and projectile loom. In all four types of loom, the filling yarns are measured and cut, thus leaving a fringe along the side. This fringe may be fused to make a selvage, if the yarns are thermoplastic, or the ends may be turned back into the cloth. These selvages are not always as usable as conventional selvages because of a tendency toward puckering that requires slitting.

Water-jet Loom

3   The water-jet loom uses a high-pressure jet of water to carry the filling yarn across the warp. It works on the principle of continuous feed and minimum tension of the filling yarns, so it can weave fabrics without barre or streaks. The filling yarn comes from a stationary package at the side of the loom, goes to a measuring drum that controls the length of each filling, and continuous by going through a guide to the water nozzle, where a jet of water carries it across through the warp shed. After the filling is carried back, it is cut off. If the fibers are thermoplastic, a hot wire is used to cut the yarn, fusing the ends so they serve as a selvage.

4   The water is removed from the loom by a suction device. Water from the jet will dissolve regular warp sizing, so one of the problems has been that of developing water-resistant sizing that can be removed easily in cloth finishing processes. The fabric is wet when it comes from the loom and must be dried (at added expense).

The loom is more compact, less noisy, and takes up less floor space than the conventional loom. It can operate at 400 to 600 picks per minute—two or three times faster than the conventional loom. Maintenance is relatively easy.

Air-jet Loom

5   Air-jet weaving is more popular because the machine costs less to purchase, install, operate, and maintain than rapier or projectile weaving machines, and the air jet can be used on a broader variety of yarns than a water jet.

6   The yarn is pulled from the supply package at a constant speed, which is regulated by the rollers, located with the measuring disk just in front of the yarn package. The measuring

disk removes a length of yarn appropriate to the width of the fabric being woven. A clamp holds the yarn in an insertion storage area, where an auxiliary air nozzle forms it into the shape of a hairpin.

7 The main nozzle begins blowing air so that the yarn is set in motion as soon as the clamp opens. The hairpin shape is stretched out as the yarn is blown into the guiding channel of the ralay nozzles along the channel. The maximum effective width for air-jet weaving machines is about 355cm. At the end of each insertion cycle the clamp closes; the yarn is beaten in, then cut, after the shed is closed. Again, some selvage-forming device is required to provide stability to the edge of the fabric.

8 The loom can operate at 320 picks per minute and is suitable for spun yarns. There are limitations in fabric width because of the diminution of the jet of air as it passes across the loom.

Rapier-type Loom

9 The rapier system operates with flexible or rigid metal arms, or rapiers, attached at both sides of the weaving area. One arm carries a pick to center of the weaving area; the arm extending from the other side grasps the pick and carries it across the remaining fabric width. Newer rapier machines are built with two distinct weaving areas for two separate fabrics. On such machines one rapier picks up the yarn from the center, between the two fabrics, and carries it across one weaving area. As it finishes laying that pick, the opposite end of the rapier picks up another yarn from the center, and the rapier moves in the other direction to lay a pick for the second weaving area, on the other half the machine.

10 Rapier looms weave more rapidly than most shuttle looms but more slowly than most projectile machines. The rapier-type loom weaves (primarily) spun yarns at 300 picks per minute. An important advantage of rapier looms is their flexibility, which permits the laying of picks of different colors. They also weave yarns of any type of fiber and can weave fabrics up to 110 inches in width without modification. This loom has found wide acceptance for use with basic cotton and wool worsted fabrics.

Projectile Loom

11 The projectile loom uses a small bullet-shaped projectile with a gripper that pulls the yarn off the supply package at the side of the weaving areas and carries it cross the shed. Only enough yarn for one pass across the width of the fabric is carried across. Projectile looms with one or more projectiles are available; the multiple-projectile type is more common. The single-projectile system picks up yarn on the supply side and carries it the entire width of the shed. After beat-up has occurred, the projectile picks up yarn from a second supply source on the other side and returns across the shed to place the next pick.

12 Multiple-projectile systems can be used in machines with a wide weaving bed, so the projectile grippers can transfer the pick across the fabric in a relay fashion. In other multiple-

projectile systems, the gripper from the first projectile picks up yarn from the supply source and moves across the shed to lay that length of yarn; then, as beat-up occurs, the projectile drops into a conveyor system that returns it to the supply side to pick up new yarn. In the meantime, the second gripper has pulled a pick to repeat the process.

13   Each pick is individually cut, so there is a continuously woven selvage like that produced by a shuttle machine. Instead, the edges are fringed. To finish them, a tucking device is used on both sides to interlace the fringe with the last few warp yarns along each edge.

(1116 words)

New Words

acid /'æsid/ *n.*　［化］酸
barre /ba:/ *n.*　纬向条花
fringe /frindʒ/ *n.*　流苏，毛边，边缘，次要，额外补贴；*vt.* 用流苏修饰，镶边
gripper /'gripə/ *n.*　夹子，片梭；夹纱器
nozzle /'nɔzl/ *n.*　喷嘴
projectile /prə'dʒektail/ *n.*　片梭
puckering /'pʌkəriŋ/ *n.*　皱纹，褶皱
selvage /'selvidʒ/ *n.*　布等的织边，镶边，布边
sizing /'saiziŋ/ *n.*　上浆；上胶；胶料；填料
slitting /'slitiŋ/ *n.*　切口，切缝，纵切，纵裂
streak /stri:k/ *n.*　纹理，条纹，斑纹，条痕，条层，色条，色线
water-resistant　抗水的，防水的

Phrases and Expressions

air-jet loom   喷气织机
picker stick   投梭棒
rapier loom   剑杆式投纬织机
shuttleless loom   无梭织机
suction device   吸水装置
water-jet loom   喷水织机

Key Sentences

1. The speed with which the shuttle is sent back and forth is limited by the mass of shuttle —

usually about 200 picks per minute.

梭子前后移动的速度受梭子本身质量的限制——通常是每分钟投梭 200 次。

2. There are limitations in fabric width because of the diminution of the jet of air as it passes across the loom.

引纬喷射气流在穿过织机的过程中有所衰减，因此织物的幅宽是有限的。

3. The loom is more compact, less noisy, and takes up less floor space than the conventional loom.

和传统织机相比，该织机结构更紧凑，噪声更小，占用的空间更少。

4. Air-jet weaving is more popular because the machine costs less to purchase, install, operate, and maintain than rapier or projectile weaving machines, and the air jet can be used on a broader variety of yarns than a water jet.

喷气织机更受欢迎，因为和剑杆织机或片梭织机相比，其购买、安装、操作和维护的成本更低，同时就可应用的纱线种类而言，又比喷水织机更广泛。

Notes

1. 无梭织机（shuttleless loom）：引纬方式是多种多样的，有剑杆、喷射（喷气、喷水）、片梭、多梭口（多相）和编织等方式。

2. 喷水织机（water-jet loom）：利用水作为引纬介质，以喷射水流对纬纱产生摩擦牵引力，将固定筒子上的纬纱引入梭口。喷水织机具有速度快、单位产量高的特点，主要适用于表面光滑的疏水性长丝化纤织物的生产。

3. 喷气织机（air-jet loom）：用喷射出的压缩气流对纬纱进行牵引，将纬纱带过梭口。喷气织机最大特点是车速快、劳动生产率高，适用于平纹和有纹路织物、细特高密织物和批量大的织物的生产。

4. 剑杆织机（rapier loom）：用刚性或挠性的剑杆头或带来夹持、导引纬纱。剑杆织机除了适宜织造平纹和有纹路织物外，其特点是换色方便，适宜多色纬织物，适用于色织、双层绒类织物、毛圈织物和装饰织物的生产。

5. 片梭织机（projectile loom）：以带夹子的小型片状梭子夹持纬纱，投射引纬。片梭织机具有引纬稳定、织物质量优、纬回丝少等优点，适用于多色纬织物、细密、厚密织物以及宽幅织物的生产。

Post-Reading Exercises

- **Reading Comprehension**

Directions: *Select the most appropriate answer for each of the following questions.*

1) What's the reason that manufacturers were devoted to seeking a way to replace the shuttle?

A. Lay the filling. B. Lay the picks.
C. Increase the speed of weaving. D. Limit the mass of shuttle.

2) Why were shuttleless looms widely adopted?
A. They were without the shuttle. B. They increased productivity.
C. They lowered noise levels. D. Both B and C.

3) How many types of shuttleless looms are referred to in the text?
A. Three. B. Four. C. Five. D. Six.

4) The advantages of water-jet looms include_____.
A. being more compact and less noisy B. taking up less floor space
C. relatively easy maintenance D. all of the above

5) Among the shuttleless looms mentioned in the text, which one is the most popular?
A. Water-jet loom. B. Air-jet loom.
C. Rapier-type loom. D. Projectile loom.

6) Which loom weaves more slowly than most projectile machines but more rapidly than most shuttle looms?
A. Water-jet loom. B. Air-jet loom.
C. Rapier-type loom. D. Projectile loom.

7) Which of the shuttleless looms permits the laying of picks of different colors?
A. Water-jet loom. B. Air-jet loom.
C. Rapier-type loom. D. Projectile loom.

- **Vocabulary**

Directions: *Complete the following sentences, using the words from the text.*

1) The water-jet loom uses a _____ jet of water to carry the filling yarn across the warp.

2) The hairpin shape is _____ _____ as the yarn is blown into the guiding channel of the relay nozzles along the channel.

3) The projectile loom uses a small bullet-shaped projectile with a gripper that pulls the yarn off the _____ _____ at the side of the weaving areas and carries it across it.

- **Translation**

Directions: *Translate the following Chinese sentences into English.*

1）厂商长久以来在寻找方法来替代旧式织布机以提高编织速度。

_____

2）梭子前后移动的速度受梭子本身质量的限制。

_____

3）织物的幅宽有一定的局限性。

4）和传统织机相比，该织机结构更紧凑，噪声更小，占用的空间更少。

5）喷气织机更受欢迎，因其购买、安装、操作和维护的成本更低廉。

Passage 3   Choosing Velvet 'Pile' Types of Fabric

1   Firstly recognize that velvet is often confused with velveteen, panne velvet and corduroy. To these names you can also add finishing processes known as devore and burnout velvet or flock and embossed velvet which can confuse even more.

2   Basically velvet, velveteen, panne velvet, and corduroy are all pile fabrics that stand up from the back of the cloth. The surface of the fabric is a series of loops which can be cut or left uncut dependant on method of manufacture or end product. An uncut pile fabric will have a pile surface whereas a cut pile fabric will have a nap surface. Pile fabrics can be made by weaving, knitting or tufting. Similar techniques are used in carpet construction, and toweling can be made from cut or uncut (loop) warp pile.

3   Velvet has a very close and dense pile. There are several methods of weaving it, but for centuries it has often been woven as a double cloth. The warp loops are formed over wire rods. Next the loops are cut during the weaving process. The double cloth is then separated and processes begin which help the bloom of the velvet to develop. After dyeing it is sheared or cropped even further to make it level. Then it is brushed so the cut threads splay out and stand up from the surface backing. Steaming gives the velvet the bloom that makes the fabric so appealing. The finished dense piled cut fabric is called velvet. People either love or hate the texture of velvet. Men often find the texture of velvet on a woman very alluring.

4   Velvet can be made from silk, rayon, nylon, polyester or cotton. I like silk, silk with viscose, and cotton velvet the most. Silk or silk viscose mixes are usually luxurious, soft and flowing and very suitable for evening wear as they drape exceptionally well. They are, though, slippery to handle and more difficult to sew than cotton velvet or velveteen, so read my tips for sewing and pressing velvet. Cotton velvet is hard wearing and tailors well for daywear. Velvet is best dry cleaned. Pile velvet, corduroy and velveteen should never be pressed in the normal way of ironing. They are all are best carefully teased and steamed with a velvet board, kettle and special pressing instructions as stated below.

Velvet Nap

5   A cut pile fabric is said to have a nap surface. This nap surface is one factor which makes construction of velvet items more difficult in both industry and at home. An item needs all the pieces to be cut in one direction. If you cut with the nap the item will feel smooth in wear as you run your hands down a garment. Cut against the nap and the colour will be deeper, richer and more luxuriant in appearance. The most important point is that every piece of a particular garment runs in the same direction.

Velveteen

6   Velveteen is mostly made from cotton and the weft pile loops are cut short. It is not so

expensive as good velvets, but quality velveteen can be quite luxurious and even hardwearing. After weaving and brushing and cropping, dye is applied by brushes which all help add the lustre and bloom peculiar to this cloth. Velveteen looks like velvet, but lies much flatter. Nonetheless all of the pieces of one complete outfit must still be cut in one direction. Velveteen has a similar effect to velvet in appearance and can make the wearer look slimmer than when wearing velvet. However velvet usually looks much more expensive. For sewing, velveteen also has a little more body so is much easier to sew. Velveteen is also easy to launder and can be washed at home and given a short spin. It dries very well in front of a gas fire or radiator which can bring up the pile.

Corduroy

7　Corduroy is made mostly from cotton. Long wefts span several warps and when the weft is cut it creates the familiar high raised lines or cords with fine backing fabric lines between. The cords lines run the length of the warp. Like velvet, corduroy should be cut in one direction only when making a garment or using heavier elephant cords in upholstery.

8　This material is used to produce casual wear such as trousers, jeans, caps and jackets. It often has other names such as corded velveteen, elephant cord, pin cord. Manchester cloth as it was produced as a Manchester cotton textile and worn originally by poorer workers in the same way that fustian was used. Manchester cloth was very good quality with dense pile but is virtually impossible to obtain today. Cotton corduroy today is often mixed with Lycra to make the fabric easier to wear and retain shape.

Panne Velvet

9　Panne velvet and panne velour are a knit velour velvet fabric in which the surface pile is directionally flattened. On occasion it comes into fashion, but is often used to portray vamps in dramatic productions as it has a cheaper look to it. It is very comfortable to wear as it is knit fabric and has incredible drape for cowl and bias cut dresses.

(893 words)

New Words

chimique /ˈʃimik/ *n.* 化学；*adj.* 化学的
cord /kɔːd/ *n.* 锁结
corduroy /ˈkɔːdərɔi/ *n.* 灯芯绒
embossed /imˈbɔst/ *adj.* 浮雕式压花的
flock /flɔk/ *n.* 植绒
haberdashery /ˈhæbəˈdæʃəri/ *n.* 男子服饰用品店
lustre /ˈlʌstə(r)/ *n.* 光彩，光泽
Lycra /ˈlaikrə/ *n.* 莱卡；人造弹性纤维品牌

nap /næp/ *vt. n.* 绒
panne /pæn/ *n.* 平绒
pile /paɪl/ *n.* 绒毛
shear /ʃɪə(r)/ *v.* 剪
tufting /'tʌftɪŋ/ *n.* 丝线法
warp /wɔ:p/ *n.* 经线
upholstery /ʌp'həʊlstəri/ *n.* 家具装饰用品
velour /və'lʊə(r)/ *n.* 天鹅绒
velveteen /ˌvelvə'ti:n/ *n.* 平绒
viscose /'vɪskəʊz/ *n.* 黏胶
weft /weft/ *n.* 纬线

Phrases and Expressions

devoré velvet (also called burnout velvet)　　　　烧花绒，烂花绒
panne velvet　　　　平绒
wire rod　　　　钢丝，钢丝筋条

Key Sentences

1. To these names you can also add finishing processes known as devore and burnout velvet or flock and embossed velvet which can confuse even more.

在上述的名称中，你还可以再加入一些和整理过程相关的名称，如烧花绒、植绒和压花绒，这些名称会让人更容易混淆。

2. The surface of the fabric is a series of loops which can be cut or left uncut dependant on method of manufacture or end product.

这种织物的表面是一系列的圈形纤维，可以割开，也可以不割开，取决于其生产方式或终端产品。

3. Silk or silk viscose mixes are usually luxurious, soft and flowing and very suitable for evening wear as they drape exceptionally well.

真丝或者真丝黏纤混纺的料子通常来说华丽、柔软、平顺，非常适合做晚装，因为它们的垂坠性非常棒。

4. A cut pile fabric is said to have a nap surface.

割过绒的面料就是我们所说的有毛绒绒表面的料子。

5. Long wefts span several warps and when the weft is cut it creates the familiar high raised lines or cords with fine backing fabric lines between.

长长的纬线与几根经线织在一起，当纬线被割断就会生成熟悉的高高立起来的线或者锁节状的线，其中间有优质的起到支撑和连接作用的线。

Notes

Devoré (also called burnout): is a fabric technique particularly used on velvets, where a mixed-fibre material undergoes a chemical process to dissolve the cellulose fibers to create a semi-transparent pattern against more solidly woven fabric. The same technique can also be applied to textiles other than velvet, such as lace or the fabrics in burnout T-shirts.

Post-Reading Exercises

- **Reading Comprehension**

Directions: *Read the passage and decide whether the following statements are true(T) or false(F) and give reasons.*

1) _____ If you know things about devore and burnout velvet or flock and embossed velvet, nothing will confuse you.

2) _____ Usually velvet, velveteen, panne velvet, and corduroy are all pile fabrics that stand up from the middle of the cloth.

3) _____ Velvet can be made from silk, rayon, nylon, polyester or cotton.

4) _____ Silk or silk viscose mixes are usually luxurious, soft and flowing and very suitable for evening wear as their cost is exceptionally high.

5) _____ Panne velvet is very comfortable to wear as it is knit fabric and has incredible drape for cowl and bias cut dresses.

- **Vocabulary**

Directions: *Complete the following sentences with the proper prepositions and adverbs.*

1) Break the chocolate _____ small pieces and melt it over a gentle heat.

2) These garments are produced _____ vast numbers.

3) I can give you an explanation _____ why I am late.

4) The surface of the fabric is a series of loops which can be cut or left uncut dependant _____ method of manufacture or end product.

5) Susan's hard working resulted _____ her getting a part in the fashion house.

- **Translation**

Directions: *Translate the following Chinese sentences into English.*

1）真丝的料子往往华丽、柔软和平顺。
_____

2）真丝或者真丝黏纤混纺的料子非常适合做晚装，因为它们的垂坠性非常棒。
_____

3）这种生产工艺依赖于其他生产过程。
_____

4）刷毛工艺可以帮助增加面料的光泽度。
_____

5）所有这种面料必须一个方向剪裁。
_____

Unit 5   Knitting Technology

PART ONE   Warm-up Activities

Cotton Textiles Touch Every Aspect of Our Lives

New Words

archaeological /ˌɑːkiəˈlɔdʒikl/ *adj.* 考古学的	exceeding /ikˈsiːdiŋ/ *n.* 超过，超越
furnishings /ˈfəːniʃiŋz/ *n.* 家具	enhance /inˈhɑːns/ *vt.* 加强，提高

Directions: *Listen to the recording and choose the most appropriate answer to each of the following questions.*

1. What does the passage mainly talk about?
   A. A kind of crop.   B. A kind of animal.
   C. A kind of fruit.   D. A kind of person.
2. It was likely that the ____ had cotton as early as 12,000 BC.
   A. Mexicans   B. Americans
   C. Egyptians   D. Australians
3. The earliest written record of Indian cotton is more than ____ old.
   A. 1,500 years   B. 2,000 years
   C. 7,000 years   D. 3,000 years

Textile Arts around the World (1)

New Words

knot /nɔt/ *v.* 打结、缠结	yarn /jɑːn/ *n.* 纱线、纺线
linen /ˈlinin/ *n.* 亚麻布	nylon /ˈnailɔn/ *n.* 尼龙
acrylic /əˈkrilik/ *n.* 丙烯酸纤维	polyester /ˌpɔliˈestə(r)/ *n.* 聚酯纤维、涤纶

Directions: *Listen to the first part of the passage twice and fill in each blank with the information you get from the recording.*

1. A textile is a piece of cloth that has been formed by _____, _____, _____

or _____ together individual pieces of fiber.

2. Yarn is a general term for long pieces of _____ fibers. Yarn can be made from _____ materials such as cotton, linen, silk and wool. Or it can be made from _____ materials such as nylon, acrylic and polyester. The paints that give color to yarn are called _____.

3. Manufacturing cloth is now a very _____ _____ process. But this was not always the case.

4. Humans probably first made textiles to meet important needs. These include textiles for _____ warm, _____ shelter, and _____ goods. But cultures around the world also developed methods of making cloth that were _____, _____, and _____.

Textile Arts around the World (2)

New Words

warp /wɔ:p/ n. 经纱	weft /weft/ n. 纬纱，织物
loom /lu:m/ n. 织布机	tapestry /'tæpəstri/ n. 挂毯
kilim /ki'li:m/ n. 绣织地毯	tribe /traib/ n. 部落
nomadic /nəu'mædik/ adj. 游牧的	geometric /ˌdʒi:ə'metrik/ adj. 几何的
piece /pi:s/ n. 补缀	Amish /'a:miʃ/ n./adj. 亚米希（人）（的）
embroidery /im'brɔidəri/ n. 刺绣	stitch /stitʃ/ n. 缝法、针法
batik /bə'ti:k/ n. 蜡染色法	

Directions: *Listen to the second part of the passage twice and choose the most appropriate answer to each of the following questions.*

1. A tapestry is a special kind of weaving method in which_____.

    A. a loom is used to weave together warp and weft threads

    B. a weaver uses the weft threads to create individual areas of color

    C. the warp threads would go in an East-West direction

    D. the weft covers the whole width of the fabric

2. Which of the following is TRUE about the seven Unicorn Tapestries in the collection of the Metropolitan Museum of Art in New York City?

    A. They are actually paintings rather than weavings because they are so detailed.

    B. These beautiful tapestries were made in the early 14th century.

    C. During the time when these tapestries were made, wealthy people used them mainly to show off their wealth.

    D. They were thought to have been designed in Paris and woven in Brussels, then part of the

Netherlands.

3. Kilims are popular in the following places except_____.
   A. North Africa   B. The Middle East
   C. Europe   D. Turkey

4. What's the main difference between kilims and those "pile" carpets?
   A. Kilims were often made by tribes that moved from place to place.
   B. Kilims are often called "Oriental" or "Persian" carpets.
   C. The surfaces of those "pile" carpets are covered with the ends of thousands of pieces of yarn.
   D. Those "pile" carpets are woven with many bright patterns and complex geometric forms.

5. What is the Amish religious group well known for?
   A. creative quilt patterns.   B. long history of making quilts.
   C. embroidery work.   D. bold use of colors.

6. Batik involves using wax to make complex patterns. In which place do people use the batik method to dye fabrics?
   A. Japan.   B. India.   C. Iran.   D. Indonesia.

PART TWO   Reading Activities

Passage 1   Modern Knitting Technology Trend: Seamless Technology

1 There has been great interest in the development of seamless knitted garments. A seamless knitted product has a number of advantages over the traditional cut and sewn product.

2 •The process of cutting and sewing is labor intensive.

3 •There is a concentration of stress where the seams are located which can jeopardize performance properties and ultimately result in premature product failure.

4 •Cutting and sewing is often manually executed, which introduces the potential for human error.

5 •The sewing process can create needle holes in the fabric which can damage the yarn.

6 •Fabric scraps produced from the cut process are discarded, resulting in fabric waste.

7 •Eliminating the cut and sew process allows for "quick-response production".

8 In addition, seams in a garment create bulkiness especially at the shoulders and underarms which can adversely affect the comfort. For these reasons, the development of seamless shaped knitted structure has been an area of interest.

9 There are a number of different machines that can produce shaped knitted products. Some of the machines totally eliminate the cut and sew process to produce a seamless garment. Other machines minimize the cut and sew processes to produce a garment with fewer seams than a traditionally made garment. Machines can be divided into three categories—each having a different level of sophistication.

10 Some machines produce shaped panels such as the front and back bodies and sleeves. After knitting, the shaped panels are sewn together to produce a garment. Other garment details such as collars and pockets are added during the sewing process. This type of process is known as shaping or full-fashion (or full-fashioned, fully fashioned).

11 Integral knitting is similar to full-fashion knitting. In integral knitting, shaped pattern pieces

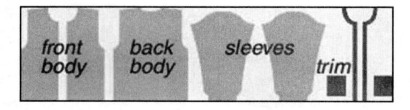

Full-fashion knitting

or panels are formed on the machine. However, unlike full-fashion knitting, trims, pockets and other details such as button holes can also be added during the knitting process. Both the full-fashion and integral knitting processes reduce the amount of cutting and sewing necessary to produce a completed garment.

12 The last and most sophisticated process is referred to as whole garment knitting, a term coined and now registered by Shima Seiki. In this process a whole garment is produced directly from the machine. No cutting or sewing is necessary. A Computer Aided Design System (CAD) is utilized to create the garment pattern. The pattern information is saved on a diskette and transferred to the knitting machine.

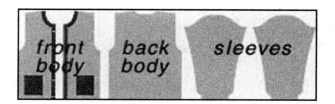 

Integral knitting   Whole garment knitting

13 To knit a sweater, three shaped tubes are knit simultaneously. A front and a back needle bed are utilized to knit the tubes. Loops are knit and transferred between the front and back beds to create shape. Three yarn carriers are used; one to knit the right sleeve, the second to knit the body, and the third to knit the left sleeve. Once knitting reaches the under arm area, the tubes are combined. The two carriers knitting the sleeves are taken out of the knitting zone. The carrier knitting the body begins to knit one tube—combining the three tubes. Garment details can be added during the knitting process.

14 Since whole garment knitting debuted more than a decade ago, there have been continuous improvements in machine design. Some of the improvements have increased productivity and design capabilities and better fabric quality.

15 Whole garment knitting machines have unique design capabilities. Although there are conventional circular knitting machines that have some of the design capabilities of whole garment machines, they are complicated to program and very expensive. One unique feature of a whole garment knitting machine is that individual stitches within a course can be controlled, allowing for increased design capabilities. Traditional circular knitting machines do not have this capability.

16 Today's whole garment machines are capable of knitting a wide variety of constructions. Everything from a sheer to bulky knit fabric can be produced. Needles can be taken out of action to knit fabrics of different weights. Machines come in different gauges, allowing for a wide variety of yarns to be used.

(682 words)

New Words

bulkiness /'bʌlkinis/ n. 庞大，笨重
capacity /kə'pæsəti/ n. 能力，容量，资格，地位，生产力
carrier /'kæriə/ n. 导纱器，横机机头
coin /kɔin/ n. 硬币，钱币；vt. 杜撰，创造，铸造（钱币）
concentration /ˌkɔnsən'treiʃən/ n. 浓度，浓缩；专心，集合
conventional /kən'venʃənəl/ adj. 传统的，符合习俗的，常见的，惯例的
course /kɔ:s/ n. 线圈横列
construction /kən'strʌkʃən/ n. 组织结构
debut /'deibju:/ vi. 初次登台；n. 初次登台，开张
discard /dis'ka:d/ n. 被抛弃的东西或人；vt. 抛弃，放弃，丢弃；vi. 放弃
eliminate /i'limineit/ vt. 消除，排除
gauge /ɡeidʒ/ n. 机号
integral /'intiɡrəl/ adj. 积分的，部分的，整体的；n. （数学）积分，部分，完整
jeopardize /'dʒepədaiz/ vt. 危害，使陷危地
loops /lu:ps/ n. 线圈；vt. 使……成环
potential /pəu'tenʃəl/ adj. 潜在的，可能的；n. 潜能，可能性，电势
panel /'pænl/ n. 衣片
scrap /skræp/ n. 碎片，残余物；adj. 废弃的，零碎的；vt. 废弃，拆毁；vi. 吵架
simultaneously /ˌsiməl'teiniəsli/ adv. 同时地
sophistication /səˌfisti'keiʃən/ n. 复杂，老于世故，有教养
stitches /'stitʃis/ n. 线圈
sweater /'swetə/ n. 毛衣
trim /trim/ n. 布边
underarm /'ʌndəra:m/ adj. 腋下的，手臂内侧的；adv. （接球时）用低手

Phrases and Expressions

button holes	纽孔
Computer Aided Design System (CAD)	计算机辅助设计系统
circular knitting machines	圆型针织机
cut and sew product	裁剪缝制而成的（服装）产品
integral knitting	整体编织
knitting machine	针织机
knitting zone	编织区域

needle bed	针床
quick-response production	快速反应生产（模式）
seamless knitted garment	无缝针织服装
Shima Seiki	日本岛精公司
whole garment knitting	无缝编织

Key Sentences

1. Some machines produce shaped panels such as the front and back bodies and sleeves.
某些针织机生产成形衣片，如前片、后片和袖片。

2. However, unlike full-fashion knitting, trims, pockets and other details such as button holes can also be added during the knitting process.
然而，与全成形针织不同的是，可以在针织工艺中加入包边、口袋及其他附件，如纽孔。

3. Loops are knit and transferred between the front and back beds to create shape.
在前后针床间进行移圈编织以成形。

4. One unique feature of a whole garment knitting machine is that individual stitches within a course can be controlled, allowing for increased design capabilities.
无缝编织机的一个特点是可以对线圈横列中的单个线圈进行控制，从而提升了设计能力。

Notes

1. 针织（knitting）：是利用织针把纱线弯成线圈，然后将线圈相互串套而成为针织物（knitted fabric）的一门纺织加工技术。根据工艺特点的不同，针织生产可分纬编（weft knitting）和经编（warp knitting）两大类。

2. 机号（gauge）：用来表明各类针织机织针的粗细和针距的大小。机号是用针床上规定长度（如 1 英寸）内所具有的针数来表示。

Post-Reading Exercises

• **Reading Comprehension**

Directions: *Read the passage and complete the following sentences.*

1) The disadvantages of traditional cutting and sewing.

　　a. The process is _____ _____.

　　b. There is a _____ where the seams are located which can _____ performance properties and result in _____ product failure.

c. The manually executed way will introduce the _____ for human error.

d. The process can create _____ _____ in the fabric which can _____ the yarn.

e. The cut process will produce _____ _____ which result in fabric waste.

f. Seams in a garment create _____ (esp. at the shoulders and underarms) which can affect _____.

For the above reasons, the development of seamless shaped knitted structure has been an area of interest.

2) There are a number of different machines that can produce shaped knitted products.

a. Some totally _____ the cut and sew process to produce a seamless garment.

b. Some _____ the cut and sew processes to produce a garment with fewer seams than a traditionally made garment. Machines can be divided into three categories according to different levels of _____.

3) There are three processes before producing a completed garment.

a. Shaping or full-fashion. Shaped panels (the front and back bodies and sleeves) are first knitted, and then are sewn together. Other garment _____ (collars and pockets) are added during the sewing process.

b. _____ knitting. Trims, pockets and other details can also be added during the knitting process.

c. Whole garment knitting. It's also the most _____ process, in which a whole garment is produced directly from the machine. No cutting or sewing is necessary. A _____ _____ _____ _____ (CADS) is used to create the garment pattern.

- **Vocabulary**

Directions: *Try to distinguish the specific meaning of the very word in each individual sentence.*

1) detail

a. *provide details for*

b. *assign to a specific task*

c. *an isolated fact that is considered separately from the whole*

d. *extended treatment of particulars*

e. *a crew of workers selected for a particular task*

____ ① The program invites the company to **detail** representatives to discuss their experience of hunting for a job.

____ ② It wasn't long before unofficial reports began to **detail** mental-health problems in the slum community.

___ ③ As the game evolves more and more *details* will be added to it.

___ ④ A *detail* was sent to remove the fallen trees in the tsunami.

___ ⑤ We still do not have final leading actors and *detail*, at present inconvenience leaks more *detailed* information.

___ ⑥ Research designed for individual surveys is explained in *detail* in the full reports.

___ ⑦ Other garment *details* such as collars and pockets are added during the sewing process.

2) potential

   a. *the inherent capacity for coming into being*

   b. *the difference in electrical charge between two points in a circuit expressed in volts*

   c. *potential possibility*

   d. *existing in possibility*

   e. *expected to become or be; in prospect*

___ ① Lifelong learning is a process to maximize individual *potential*.

___ ② All vitamins have the *potential* to be toxic if you ingest too many of them.

___ ③ Many investors are nervous about the *potential* for inflation.

___ ④ There are plenty of *potential* sources of friction between the two super powers.

___ ⑤ The company mailed the catalog out to all *potential* customers.

___ ⑥ The relation between *potential* and electric field intensity is an important content in electromagnetics.

___ ⑦ Cutting and sewing is often manually executed which introduces the *potential* for human error.

- **Translation**

Directions: *Translate the following sentences into English, using the expressions in brackets.*

1) 裁剪和缝纫往往是手工操作的，可能会产生人为的失误。（execute, introduce）

2) 自从十多年前无缝编织首次出现以来，机器设计方面已取得了不断的改进。(debut)

3) 今天，无缝编织机业已能制造各种样式的服装，包括透明的薄织物服装以及厚重的服装。（be capable of, a wide variety of）

4) 最后也是最复杂的工艺是整体服装编织。（sophisticated process）

5) 裁剪工序中产生的织物废料被丢弃，产生织物损耗。（cut process）

Passage 2　The Origins of the Knitting Machine

1　Hand-knitting was well established in the Middle East a thousand years before the invention of the knitting machine. The invention and development of that first knitting "frame", however, took place in England, and was preceded by the introduction of hand-knitting to this country. The art was introduced to Europe from the Middle East by trade and conquest, and gradually spread through the continent, finally arriving in the British Isles. It is thought probable that Spain was the first European country to acquire the art when the Arabs gained control of the Iberian Peninsula at the beginning of the 8th century.

2　References to hand knitting occur in the literature of several European countries as the art passed gradually through the different regions. Writings infer that Pope Innocent Ⅳ, who died in 1254, had a pair of hand-knitted silk gloves; a biography, published in Germany, indicates that hand knitting was practiced there at the turn of the 14th century; it was reported that Henry Ⅱ of France wore hand-knitted silk stockings when he married Catherine de Medici in 1533.

3　The earliest known reference to knitting in England occurs in the Accounts of the Chapter of *the Collegiate Church of St. Peter and St. Wilfred at Ripon* for the year 1452; a "knyt gyrdyir" is mentioned. By 1488, knitting was well established, since it is referred to in an Act of Parliament of that year:

4　"The price of felted hats to be 1s. 8d. and the price of knytted wollen caps should be 2s. 8d."

5　During the second half of the 15th century and the first half of the 16th century, we find numerous other references to hosiery and knitting. Before this time men wore long hose made of woven cloth, cut and seamed to shape, it was noted.

6　The inventor of the "frame-work knitting machine", as it is called in contemporary accounts, was a clergyman, the Reverend William Lee. He was born in Nottinghamshire, near to what is now the City of Nottingham. Two villages claim to have had the honor of his birth—Calverton and Woodborough; we cannot be sure which one, because the parish records of both villages are incomplete. It is certain, however, that Lee, after graduating with the degree of Master of Arts from Cambridge University, became Curate of Calverton. It seems extraordinary that a scholar and cleric, with no mechanical background or training, was able, through skill and perseverance, to produce the intricate knitting frame and lay the foundations for the modern hosiery trade.

7　A number of stories have been told to account for a man of Lee's background turning inventor; three main ones seem to persist.

8　The first story says that Lee, while a student, married secretly, and that his wife produced a child very early in the marriage. His wife attempted to support the family by knitting stockings by hand; Lee, in despair, was watching his wife knit one day when he conceived the idea of inventing a machine to do the work.

9 The second story holds that Lee was deeply in love with a girl who, whenever he went to visit her, seemed more interested in her knitting than she was in him. This developed in Lee an aversion for hand-knitting, and made him keen to be instrumental in destroying it as an occupation (Even though he did invent a knitting machine, he failed to destroy handknitting).

10 The third version has it that Lee fell in love with a woman who had become rich by employing other girls as handknitters. She rejected his proposal of marriage, and Lee vowed he would create a machine which would destroy the business of the woman who had spurned him.

11 We have no means of knowing which of these stories (if any) is true, but it would seem that a woman was involved in some way.

12 What we do know, however, is that Lee invented a machine at Calverton, and the date which is generally accepted is 1589. Lee's first machine had eight needles to the inch, and could produce a coarse fabric. After the first development, William was joined by his brother James, and they determined to secure protection and financial assistance under the "Patent" system, which had begun in 1561. During the first twenty years of the system, thirteen patents had been granted for mechanical inventions, and Lee's device seemed very suitable for recognition. Lee was fortunate in securing a powerful ally in Lord Hunsdon, and there seemed every chance of success.

13 William Lee and his brother set up a workshop in London, and demonstrated the machine before Queen Elizabeth I. She rejected the application for a patent on two grounds: the stocking produced was coarse and sleazy and made from wool, whereas Elizabeth always wore hand knitted silk hose; and such a machine would, she thought, destroy the livelihood of the thousands of "stockingers" who knitted by hand. The Queen, however, implied that if Lee could produce finer stockings from silk, she would reconsider the matter. So Lee set to work to produce a finer machine, with more needles to the inch; what we should nowadays call a 'finer gauge' machine. It was only in the year 1598, however, that Lee was able to produce the first-ever fully-fashioned silk stocking. He presented a pair to the Queen, which she accepted and praised. She did not, however, grant a patent; Lee was still waiting for that when the Queen died in 1603.

14 William Lee attempted to interest her successor, James I, in his invention, but without success. By this time, William and James Lee had built nine frames, and had a number of assistants. They were able to produce and sell enough stockings to make a precarious living.

(965 words)

New Words

aversion /ə'və:ʃən/ *n.* 讨厌，讨厌的人
biography /bai'ɔgrəfi/ *n.* 传记，档案，个人简介
claim /kleim/ *vi.* 提出要求； *vt.* 要求，声称，认领；

n. 要求，声称，索赔，断言，值得

clergyman /ˈklə:dʒimən/ *n.* 牧师，教士

cleric /ˈklerik/ *n.* 牧师，教士；*adj.* 牧师的，教士的

conceive /kənˈsi:v/ *vi.* 怀孕，设想，考虑；*vt.* 怀孕，构思，以为，持有

coarse /kɔ:s/ *adj.* 粗俗的，粗糙的

contemporary /kənˈtempərəri/ *adj.* 当代的，同时代的；*n.* 同时代的人，同时期的人

curate /ˈkjurit/ *n.* 助理牧师，副牧师

frame /freim/ *n.* 框架，结构；*adj.* 有木架的，有框架的；*vt.* 设计、建造、陷害

hose /həuz/ *n.* 软管，长统袜，男性穿的紧身裤；*vt.* 用软管浇水，痛打

hosiery /ˈhəuziəri/ *n.* 针织品，袜类

indicate /ˈindikeit/ *vt.* 表明，指示，预示，象征

intricate /ˈintrikət/ *adj.* 复杂的，错综的

needle /ˈni:dl/ *n.* 织针

parish /ˈpæriʃ/ *n.* 教区

perseverance /ˌpə:si:ˈviərəns/ *n.* 坚持不懈，不屈不挠

precede /pri:ˈsi:d/ *vt.* 领先，优于，在……之前；*vi.* 领先，在前面

precarious /priˈkɛəriəs/ *adj.* 危险的，不稳定的

present /ˈprezənt, priˈzent/ *vi.* 举枪瞄准；*adj.* 现在的，出席的；*n.* 现在，礼物，瞄准

seam /si:m/ *n.* 缝，接缝；*vt.* 缝合，结合；*vi.* 裂开，产生裂缝

stocking /ˈstɔkiŋ/ *n.* 长筒丝袜，丝袜

Phrases and Expressions

account for	对……负有责任，对……做出解释
be granted for	给予
hand knitting	手工编织
hosiery trade	针织业
Iberian Peninsula	伊比利亚半岛
knitting machine	针织机
knitting frame	针织机
lay the foundations for...	为……奠定基础
parish record	教区记录

Key Sentences

1. Lee's first machine had eight needles to the inch and could produce a coarse fabric.

李的第一台针织机每英寸有八根织针，可以生产风格粗犷的织物。

2. So Lee set to work to produce a finer machine with more needles to the inch; what we should nowadays call a 'finer gauge' machine.

为此，李着手制作针距更小的针织机，该机每英寸内有更多织针，我们现在称之为"细针距"针织机。

Notes

针织机（knitting machine）：利用织针把纱线编织成针织物的机器称为针织机。针织机按工艺类别可分为纬编针织机与经编针织机；按针床形式可分为平型针织机与圆型针织机。

Post-Reading Exercises

- **Reading Comprehension**

Directions: *Read the passage and complete the following sentences.*

1) Hand-knitting

   a. It was well established in the _____ _____, a thousand years before the invention of the knitting machine.

   b. It was introduced to _____ from the Middle East by _____ and _____, and gradually spread through the _____, finally arriving in the _____ _____.

   c. _____ probably was the first European country to acquire the art when the Arabs gained control of the Iberian Peninsula at the beginning of the 8th century.

2) Knitting Machine

   d. The invention and development of the first knitting frame took place in _____.

   e. Its inventor was a _____, the Reverend William Lee, who laid the foundations for the modern hosiery trade.

   f. A workshop was set up in London, Lee _____ the machine before Queen Elizabeth, but it was rejected due to the concerns over the destruction of the _____ of the "stockingers".

   g. In 1598, the first-ever fully-fashioned silk stocking was produced by a "_____ _____" machine.

   h. The machine still did not interest James I, the _____ of the Queen. By this time, Lee had built nine frames and the group was able to produce enough stockings to make a _____ living.

- **Vocabulary**

Directions: *Comprehend the very word in the individual sentences.*

1) contemporary

   a. The artist Michelangelo often stirred up the opposition among the ***contemporary*** artists of his day.

   b. Beethoven was a ***contemporary*** of Napoleon.

2) conceive

   a. She urges these patients who are considering pregnancy to lose a few pounds before they ***conceive***.

   b. I ***conceive*** your answer would be faint for the rain falling, while it still sounds so clear.

3) present

   a. The Committee will ***present*** its final observations and conclusions in two weeks.

   b. Then, have you received an award at your ***present*** company?

4) claim

   a. Did you ***claim*** the insurance after your car accident?

   b. Most economists would ***claim*** that the project has been a success.

   c. He resigned his ***claim*** to the copyright.

- **Translation**

Directions: *Translate the following sentences into English, using the expressions in brackets.*

1）第一台纺织"架"是手工针织被引进到英格兰之后才诞生并发展起来的。（be preceded by）

2）我们无意纠结这些故事的真假，但是某种程度上这都涉及了一项针织技术的秘密。（have no means of...but ...；is involved in ...）

3）大约于1589年李发明的针织机每英寸有八根织针，可以生产粗犷的织物。（produce）

4）他决心获得始于1561年的"专利"系统的保护以及经济援助。（secure）

5）作为一名没有机械背景以及训练的学者和教师，李通过技术和坚持不懈的努力最终生产出了复杂的针织机，并且为现代针织品贸易奠定了基础。（lay the foundation for ...）

Passage 3 A More Casual Style of Men's Tie

By Simon Brooke

1 It's one of the toughest aspects of the end of summer: along with early starts, the commute between office and home (rather than hotel to beach), and an engorged email in-tray. Back to work means back to stiff and square office gear. Happily, there is something new on the men's wear horizon that can help bridge the gap between laissez faire and white collar. Say hello to the knitted tie.

2 An under-the-radar trend for a few seasons now, this autumn the knitted tie has hit many a mainstream men's wear collection. "Men have been looking for something more casual with neckwear for quite some time now," says Ritchie Charlton, managing director of Savile Row tailor Douglas Hayward.

3 Complementing the trend for slimmer tailoring and narrow lapels, the knitted tie has a softer and more retro look. Andrew Roberts, 40, a Hayward client who works as communications director for Mercedes—Benz, prefers them for their versatility: "They can be dressed up for the evening or down for meetings during the day," he says.

Mark Baxter, head of men's wear at Austin Reed and Viyella, agrees: "The knitted tie has a great versatility that helps create an edge to a formal outfit. We're seeing more and more men wearing blazers with jeans or chinos and the knitted tie. It's making its way back as an homage to sharp 1960s styling, and it's fast becoming a staple tie for men's wear."

4 To wit: Paul Smith is going for simple stripes with a gradient colour palette (£69); Austin Reed tends towards classic elegance with plain navy or black-and-grey stripes (prices from £40); Sonia Spencer, normally known for her cufflinks, has introduced her first knit tie collection (prices from £55) inspired by Pantone, the professional colour company, with colours including lipstick red and prism violet; and Brooks Brothers does its usual preppy/smart look with muted colours such as burnt orange or baby blue and ochre in widely spaced contrasting stripes (prices from £59).

5 Marc Shrimpling, 26, a competition lawyer, became a knitted tie regular when he moved from London to Bristol, and changed his personal style accordingly. "I worked for a big city law firm that was very conservative and dull, so I always wore very serious ties. But when I moved to Bristol I started wearing some old 1980s-style knitted ties that I'd bought in a second hand shop," he says. "They work well because people-even lawyers-are more relaxed and less stuffy down here, I've found. It's a bit like wearing brown shoes."

6 Not that knitted ties are a poor relation of more formal versions. As Michael Drake of the eponymous men's wear label points out, as much work can go into a knit tie as a conventional one. "Ours start with quality raw silk which is spun, dyed, and finished in Switzerland in a way that assures the cri de la soie, the crunch of the silk, when squeezed." The Drake collection ranges from minimalist grey-on-black to contrasting coloured spots on bright backgrounds such as blue and navy

or olive green and black (prices from £85). "The yarn is knitted on hundred-year-old frames," he says, "which are slower than modern machinery but ensure a superior quality, then sewn down the centre back and put on a mould for 24 hours to achieve the desired finished shape."

7 Italian tailor Rubinacci stocks plain and striped knitted ties (prices from £60), mostly in silk, but also in cotton for summer and in wool for winter and, according to Mariano Rubinacci, "Such ties are fashionable because they're elegant, understated and, at the same time, a little bit retro, with a reference to the school ties in the 1950s."

8 Rupert Phelps, a 40-year-old banker, is a fan. "A traditional City wardrobe should include a very dark blue knitted silk tie, preferably with a horizontal rather than fishtail end, and quite thin," he says. "It's an essential bit of kit in that it's understated and goes with pretty much every cloth you could have shirts or suits made from."

9 Philip Kingsley, an 80-year-old haircare specialist, is also keen. "Knitted ties have the appeal that they can be worn in a formal way with business suits, or more informal as with a blazer or sports jacket," he says. "A well produced, quality knitted tie is elegant and never dated. And of course, they never crease. I don't even know how many I have now."

10 There are limitations, however. Simon Gotelee, 45, who consults for a firm of solicitors says, "They're hopeless with a formal City suit, and the bad vibe is geography teachers in moss or fawn-coloured wool ties, which always had a frayed knot and fluffy straggly bits at the bottom. But they can be good with a blazer or cotton jacket due to their informality. They're the only ties that work with a horizontal stripe, especially the thick knitted ones (often silk, not wool) with a square end."

11 Back at tailor Douglas Hayward, Ritchie Charlton's advice is to keep the background simple. "Knitted ties work well with solid blocks of colour such as grey or black, but brighter, simpler patterns also look better here than they often do with traditional fine silk ties."

(866 words)

New Words

blazer /'bleɪzə(r)/ *n.*　法兰绒运动上衣
chino /'tʃiːnəʊ/ *n.*　斜纹棉布，斜纹棉布裤；*adj.*　用斜纹棉布做成的
conservative /kən'sɜːvətɪv/ *adj.*　保守的；（式样等）不时新的
conventional /kən'venʃənl/ *adj.*　传统的；习用的，平常的；依照惯例的；约定的
crease /kriːs/ *vt.*　使有皱褶，弄皱；表面被子弹擦伤或击伤；*vi.*　起皱
cufflink /'kʌflɪŋk/ *n.*　（衬衣的）袖扣
engorged /en'ɡɔːdʒd/ *adj.*　塞得满满的，过饱的
eponymous /ɪ'pɒnɪməs/ *adj.*　齐名的；使得名的
fawn-coloured /'fɔːn'kʌləd/ *adj.*　淡黄褐色的

frayed /freɪd/ *adj.* 磨损的

gear /gɪə(r)/ *n.* 衣服；齿轮；排挡；传动装置

gradient /ˈɡreɪdiənt/ *adj.* 倾斜的

homage /ˈhɒmɪdʒ/ *n.* 致敬

horizontal /ˌhɒrɪˈzɒntl/ *adj.* 水平的

in-tray /ˈɪntreɪ/ *n.* （办公室使用的）公文格，收文篮

lapel /ləˈpel/ *n.* （西服上衣或夹克的）翻领

mainstream /ˈmeɪnstriːm/ *n.* （思想或行为的）主流；主要倾向，主要趋势

minimalist /ˈmɪnɪməlɪst/ *n.* 极简抽象派艺术家，极简派音乐家；保守派；
　　　　　　adj. 极简抽象艺术的，极简抽象风格的；最低必须限度的

moss /mɒs/ *n.* 苔藓；藓沼；泥炭沼

mould /məʊld/ *n.* 铸模；模型

muted /ˈmjuːtɪd/ *adj.* （颜色）柔和的，不鲜艳的

ochre /ˈəʊkə(r)/ *n.* 赭石；赭色，黄褐色

palette /ˈpælət/ *n.* 调色板，颜料

preppy /ˈprepi/ *adj.* 预备学校学生的，（尤指在衣着、举止等方面）像预备学校学生的

retro /ˈretrəʊ/ *adj.* 复古的；怀旧的；*n.* （服装式样等）重新流行；复旧，怀旧

solicitor /səˈlɪsɪtə(r)/ *n.* 咨询律师，诉状律师，事务律师（英国），法务官（美国）

straggly /ˈstræɡli/ *adj.* 脱离行列的；落后的；蔓延的；散乱的

staple /ˈsteɪpl/ *adj.* 最基本的，最重要的

stripe /straɪp/ *n.* 条纹

stuffy /ˈstʌfi/ *adj.* 闷热的，不通气的；古板的，保守的；枯燥无味的；一本正经的

under-the-radar *adj.* 不引人注目的，低调的

understated /ˌʌndəˈsteɪtɪd/ *adj.* 朴素的，简朴的；轻描淡写的；有节制的；低调的

versatility /ˌvɜːsəˈtɪləti/ *n.* 多功能性；多才多艺；用途广泛

vibe /vaɪb/ *n.* 感应，感觉

wardrobe /ˈwɔːdrəʊb/ *n.* 衣柜，衣橱；藏衣室；（个人，戏团的）全部服装；行头

Phrases and Expressions

office gear	办公室着装
laissez faire	（法语）放任，指休闲风格
knitted tie	针织领带
dressed up	（盛装）打扮
formal outfit	正装
burnt orange	焦橙色

prism violet	炫目紫
the cri de la soie	（法语）与"the crunch of the silk"同意，指丝绸受到挤压时发出的嘎吱声
weft-knitted fabric	纬编针织品，纬编织物

Key Sentences

1. An under-the-radar trend for a few seasons now, this autumn the knitted tie has hit many a mainstream men's wear collection.

如今，低调而不张扬的潮流已经延续了好几季，这个秋季针织领带影响了众多主流男士服装系列。

2. Not that knitted ties are a poor relation of more formal versions.

针织领带与较为正式的服装并非格格不入。

3. Ours start with quality raw silk which is spun, dyed, and finished in Switzerland in a way that assures the cri de la soie, the crunch of the silk, when squeezed.

我们从高品质生丝开始，纺染和最后加工都在瑞士完成，以便在受到挤压时确保艺术效果。

4. It's an essential bit of kit in that it's understated and goes with pretty much every cloth you could have shirts or suits made from.

这是服装配套的必备之品，因为它非常素雅，几乎能与衬衫或西装的任何布料搭配。

Notes

1. Austin Reed 是英国著名服装品牌，创立于 1900 年，Austin Leonard Reed 在伦敦 Fenchurch 街开设第一家私人定制店，由男士服装定制店逐渐发展成为经典的英国服装品牌。该品牌风格沉稳儒雅，面料天然优质、做工精致考究，设计简洁流畅，深受英国王室、贵族和大众的喜爱。

2. Viyella 是英国知名时装品牌，Viyella 最初是英国某公司于 1893 年开发的一种新型织物，成分为 55% 美利奴细羊毛、45% 棉，1894 年在英国注册商标"维耶勒法兰绒"，是世界上第一个拥有商标的织物，1907 年在美国注册，逐渐发展为目前的时装品牌，主营服装和家居用品。

Post-Reading Exercises

- **Reading Comprehension**

Directions: *Read the passage and decide whether the following statements are true(T) or* *false(F)* and give reasons.

1) _____ The knitted tie is contradictive with the trend for slimmer tailoring and narrow lapels.

2) _____ People may feel the knitted tie rather retro, similar to 1960s style.

3) _____ It's considered that the knitted tie is of good quality, but dull for lack of variety.

4) _____ Knitted ties are more suitable for formal business suits.

5) _____ Professionals like geography teachers love wearing knitted ties.

- **Vocabulary**

Directions: *Complete the following sentences with the words given in the box.*

understated	under the radar（低调神秘地）	retro（复古的）
minimalist	conservative	conventional

1) The _____ decoration of the apartment is an homage to the 1980s style.

2) The opposition _____ Party put a different interpretation on the figures.

3) He's a real English gentleman with his typically _____ humour.

4) In the living room, the _____ theme continues without any additional furniture or flashy colours.

5) Habits are learnt by repetition and so they can sneak in _____ _____ _____.

6) Alternative treatments can provide a useful back-up to _____ _____ treatment.

- **Translation**

Directions: *Translate the following Chinese sentences into English.*

1）给图片配上一些合适的家具，唤起了一种年代感。（complement, evoke）

_____

2）丰富的产品组合为他们公司增加了竞争中的优势。（versatility, edge）

_____

3）她比较喜欢低调的极简主义风格的家具。（go for, minimalist）

_____

4）当地酒吧的一些常客创建了基金用来帮助酗酒者。（regular）

_____

5）可供选择的颜色非常多，有淡蓝、橄榄绿、焦橙色，还有炫目紫。（baby blue, olive green, burnt orange, prism violet）

_____

Unit 6   Fashion Designers

PART ONE   Warm-up Activities

Ingenuity of Fashion

New Words

utilitarian /juːˌtɪlɪˈteəriən/ *adj.* 功利的   zeitgeist /ˈzaɪtɡaɪst/ *n.* 时代精神，时代潮流

Directions: *Listen to the passage and fill in each blank with the information you get from the recording.*

In the fashion industry, there's very little intellectual property protection. People have trademark protection, but no copyright protection and no 1. _____ protection to speak of. All they have, really, is trademark protection, and so it means that anybody could copy any 2. _____ on any person in this room and sell it as their own design. The only thing that they can't copy is the actual trademark label within that piece of 3. _____.

What I'm going to argue toady is that because there's no 4. _____ protection in the fashion industry, fashion designers have actually been able to elevate utilitarian design, things to cover our 5. _____ bodies, into something that we consider art. Because there's no copyright protection in this industry, there's a very open and 6. _____ ecology of creativity. Unlike their creative brothers and sisters, who are sculptors or photographers or 7. _____ or musicians, fashion designers can sample from all their 8. _____ designs. They can take any element from any garment from the history of fashion and incorporate it into their own design. They're also notorious for riffing off of the zeitgeist. And here, I 9. _____ they were influenced by the 10. _____ in Avatar. Maybe just a little can't copyright a costume either. Now, fashion designers have the broadest palette imaginable in this creative industry.

So, one of the 11. _____ side effects of having a culture of copying, which is really what it is, is the 12. _____ of trends. People think this is a magical thing. How does it happen? Well, it's because it's 13. _____ for people to copy one another. Some people believe that there are a few people at the top of the fashion food chain who sort of dictate to us what we're all going to wear, but if you talk to any designer at any level, including these high-end designers, they always say their main 14. _____ comes form the street: where people like you and me remix and match our own fashion looks. And that's where they really get a lot of their creative inspiration, so it's both a 15. _____ and a bottom-up kind of industry.

Match the Garments with the Appropriate Accessories

New Words

outfit /ˈautfit/ n.　全套服装
garment /ˈgɑːmənt/ n.　（一件）衣服；服装，衣着
accessory /əkˈsesəri/ n.　服装辅件，配饰

Directions: *Find out how well you can put together an outfit by matching the garments with the appropriate accessories.*

1) A flippy flower-print skirt and a tank top（印花裙和无袖背心）	A) A Panama hat[④] and a false mustache（巴拿马帽子和假胡子）
2) A three-piece seersucker suit（三件套绉纹薄纱西服）	B) Fuzzy yellow slippers and a cup of warm milk（黄色毛绒拖鞋和一杯温牛奶）
3) Cargo pants[①] and a polo shirt[②]（工装裤和POLO衫）	C) A crocheted hoodie and jelly flip-flops[⑤]（钩织连帽衫和透明橡胶平底人字拖鞋）
4) A floor-length bias-cut gown with spaghetti straps[③]（吊带斜裁拖地礼服）	D) A faux fur wrap and strappy silver high-heeled sandals（人造皮草披肩和银色高跟绑带凉鞋）
5) Cropped jeans and an embroidered V-neck tunic（七分牛仔裤和刺绣V领束身上衣）	E) Retro sneakers[⑥] and wraparound Sunglasses[⑦]（复古运动鞋和包围式墨镜）
6) Flannel pajamas with rubber duckies all over them（布满橡胶鸭子图案的法兰绒睡衣）	F) Bangle bracelets and espadrille shoes[⑧]（手镯和帆布便鞋）

Unit 6　Fashion Designers | 113

Reference: Illustrations for some items:

PART TWO   Reading Activities

Passage 1   The Occupation: Fashion Designers

1 Fashion designers help create the billions of dresses, suits, shoes, and other clothing and accessories purchased every year by consumers. Designers study fashion trends, sketch designs of clothing and accesso-ries, select colors and fabrics, and oversee the final production of their designs. ***Clothing designers*** create and help produce men's, women's, and children's apparel, including casual wear, suits, sportswear, formalwear, outerwear, maternity, and intimate apparel. ***Footwear designers*** help create and produce different styles of shoes and boots. ***Accessory designers*** help create and produce items such as handbags, belts, scarves, hats, hosiery, and eyewear, which add the finishing touches to an outfit. Some fashion designers specialize in clothing, footwear, or accessory design, but others create designs in all three fashion categories.

2 The design process from initial design concept to final production takes between 18 and 24 months. The first step in creating a design is researching current fashion and making predictions of future trends. Some designers conduct their own research, while others rely on trend reports published by fashion industry trade groups. Trend reports indicate what styles, colours, and fabrics will be popular for a particular season in the future. Textile manufacturers use these trend reports

to begin designing fabrics and patterns while fashion designers begin to sketch preliminary designs. Designers then visit manufacturers or trade shows to procure samples of fabrics and decide which fabrics to use with which designs.

3 Once designs and fabrics are chosen, a prototype of the article using cheaper materials is created and then tried on a model to see what adjustments to the design need to be made. This also helps designers to narrow their choices of designs to offer for sale. After the final adjustments and selections have been made, samples of

the article using the actual materials are sewn and then marketed to clothing retailers. Many designs are shown at fashion and trade shows a few times a year. Retailers at the shows place orders for certain items, which are then manufactured and distributed to stores.

4 Computer-aided design (CAD) is increasingly being used in the fashion design industry. Although most designers initially sketch designs by hand, a growing number also translate these hand sketches to the computer. CAD allows designers to view designs of clothing on virtual models and in various colors and shapes, thus saving time by requiring fewer adjustments of prototypes and samples later.

5 Depending on the size of their design firm and their experience, fashion designers may have varying levels of involvement in different aspects of design and production. In large design firms, fashion designers often are the lead designers who are responsible for creating the designs, choosing the colors and fabrics, and overseeing technical designers who turn the designs into a final product. They are responsible for creating the prototypes and patterns, and work with the manufacturers and suppliers during the production stages. Large design houses also employ their own patternmakers, tailors, and sewers who create the master patterns for the design, and sew the prototypes and samples. Designers working in small firms, or those new to the job, usually perform most of the technical, patternmaking, and sewing tasks, in addition to designing the clothing.

6 Fashion designers working for apparel wholesalers or manufacturers create designs for the mass market. These designs are manufactured in various sizes and colors. A small number of high-fashion (*haute couture*) designers are self-employed and create custom designs for individual clients, usually at very high prices. Other high-fashion designers sell their designs in their own retail stores or cater to specialty stores  or high-fashion department stores. These designers create a mixture of original garments and those that follow established fashion trends.

7 Some fashion designers specialize in costume design for performing arts, motion picture, and television productions. The work of costume designers is similar to other fashion designers. Costume designers, however, perform extensive research on the styles worn during the period in which the performance takes place, or they work with directors to select and create appropriate attire. They make sketches of designs, select fabric and other materials, and oversee the production of the costumes. They also must stay within the costume budget for the particular production item.

Work environment

8   Fashion designers employed by manufacturing establishments, wholesalers, or design firms generally work regular hours in well-lighted and comfortable settings. Designers who freelance generally work on contract, or by the job. They frequently adjust their workday to suit their clients'  schedules and deadlines, meeting with the clients during evenings or weekends when necessary. Freelance designers tend to work longer hours and in smaller, more congested environments, and are under pressure to please clients and to find new ones in order to maintain a steady income. Regardless of their work setting, all fashion designers occasionally work long hours to meet production deadlines or prepare for fashion shows.

9   The global nature of the fashion business requires constant communication with suppliers, manufacturers, and customers all over the United States and the world. Most fashion designers travel several times a year to trade and fashion shows to learn about the latest fashion trends. Designers also may travel frequently to meet with fabric and materials suppliers and with manufacturers who produce the final apparel products.

Education and training

10   In fashion design, employers usually seek individuals with a 2-year or 4-year degree who are knowledgeable about textiles, fabrics, ornamentation, and fashion trends.

11   Fashion designers typically need an associate's or a bachelor's degree in fashion design. Some fashion designers also combine a fashion design degree with a business, marketing, or fashion merchandising degree, especially those who want to run their own business or retail store. Basic coursework includes color, textiles, sewing and tailoring, pattern making, fashion history, computer-aided design (CAD), and design of different types of clothing such as menswear or footwear. Coursework in human anatomy, mathematics, and psychology also is useful.

12   The National Association of Schools of Art and Design accredits approximately 300  postsecondary institutions with programs in art and design. Most of these schools award degrees in fashion design. Many schools do not allow formal entry into a program until a student has successfully completed basic art and design courses. Applicants usually have to submit sketches and other examples of their artistic ability.

13   Aspiring fashion designers can learn

these necessary skills through internships with design or manufacturing firms. Some designers also gain valuable experience working in retail stores, as personal stylists, or as custom tailors. Such experience can help designers gain sales and marketing skills while learning what styles and fabrics look good on different people.

14 Designers also can gain exposure to potential employers by entering their designs in student or amateur contests. Because of the global nature of the fashion industry, experience in one of the international fashion centers, such as Milan or Paris, can be useful.

Other qualifications

15 Designers must have a strong sense of the aesthetic—an eye for color and detail, a sense of balance and proportion, and an appreciation for beauty. Fashion designers also need excellent communication and problem-solving skills. Despite the advancement of computer-aided design, sketching ability remains an important advantage in fashion design. A good portfolio—a collection of a person's best work—often is the deciding factor in getting a job.

16 In addition to creativity, fashion designers also need to have sewing and patternmaking skills, even if they do not perform these tasks themselves. Designers need to be able to understand these skills so they can give proper instruction in how the garment should be constructed. Fashion designers also need strong sales and presentation skills to persuade clients to purchase their designs. Good teamwork and communication skills also are necessary because increasingly the business requires constant contact with suppliers, manufacturers, and buyers around the world.

Advancement

17 Beginning fashion designers usually start out as pattern makers or sketching assistants for more experienced designers before advancing to higher level positions. Experienced designers may advance to chief designer, design department head, or another supervisory position. Some designers may start their own design company, or sell their designs in their own retail stores. A few of the most successful designers can work for high-fashion design houses that offer personalized design services to wealthy clients.

(1388 words)

New Words

accredit /əˈkredit/ vt.　信任，认可

adjustment /ə'dʒʌstmənt/ n. 调节，调整
apparel /ə'perəl/ n. 衣服，服装，衣着
amateur /'æmətə/ n. 业余从事者；外行人；爱好者；adj. 业余的；外行的
anatomy /ə'nætəmi/ n. 解剖，解剖学
article /'ɑːtɪkl/ n. （物品的）一件，物品；商品
attire /ə'taɪə/ n. 服装，衣着，盛装
congested /kən'dʒestəd/ adj. 拥挤的，堵塞的
costume /'kɔstjuːm/ n. 戏服，民族服装，装束
establishment /ɪ'stæblɪʃmənt/ n. 建立的机构；公司；会团；学校；机关；企业
esthetic /iːs'θetɪk/ adj. 美的，美学的；美感的
eyewear /'aɪˌweə/ n. 眼镜
fabric /'fæbrɪk/ n. 织物，织品；布料
freelance /'friːlæns/ vi. 当自由作家（或演员等）；
    n. （不受雇于人的）自由作家（或演员等）；
    adj. 自由作家（或演员等）的；独立的；
    adv. 作为自由作家；独立地
hosiery /'həʊzɪəri/ n. （总称）袜子
internship /'ɪntɜːnʃɪp/ n. 实习生的职位；实习期
maternity /mə'tɜːnəti/ n. 孕妇装
ornamentation /ˌɔːnəmen'teɪʃn/ n. 装饰品，装饰
outerwear /'aʊtəˌweə/ n. （总称）外衣，外套
oversee /ˌəʊvə'siː/ vt. 监视；监督；管理；看管
patternmaker /'pætən'meɪkə/ n. 打样师；制模师
preliminary /prɪ'lɪmɪnri/ adj. 初步的，预备的；n. 初步，开端，预备；预考；预赛
procure /prə'kjʊə/ vt. （努力）取得，获得；采办；为……获得
portfolio /pɔːt'fəʊlɪəʊ/ n. （艺术家等的）代表作选辑；文件夹，卷宗夹
prototype /'prəʊtətaɪp/ n. 原型，样板；标准，模范
scarf /skɑːf/ n. 围巾，头巾
sew /səʊ/ vt. 缝，缝合，缝纫
sketch /sketʃ/ vt. 画草图；草拟；n. 速写，素描；略图，草图，粗样，草稿
suit /sjuːt; suːt/ n. （一套）衣服
supplier /sə'plaɪə/ n. 供应者，供货商
touch /tʌtʃ/ n. 装点，润色
varying /'veərɪŋ/ adj. 变化的，不同的
wholesaler /'həʊlˌseɪlə/ n. 批发商

Phrases and Expressions

CAD (Computer-Aided Design)	计算机辅助设计
cater to	迎合，满足
custom design	订制设计，量身订制
custom tailor	制作定制服装的高级裁缝
design house	设计工作室，设计公司
high-fashion / haute couture	高级订制时装
intimate apparel	贴身衣服，内衣
mass market	大众市场
master pattern	原始样板
personal stylist	私人造型师
postsecondary institutions	泛指高中以上的学院，包括本科和大专学院
specialty store	专卖店

Key Sentences

1. Clothing designers create and help produce men's, women's, and children's apparel, including casual wear, suits, sportswear, formal wear, outerwear, maternity, and intimate apparel.

服装设计师设计和帮助生产男装、女装和童装，包括休闲装、套装、运动装、正装、外套、孕妇装和内衣。

2. Once designs and fabrics are chosen, a prototype of the article using cheaper materials is created and then tried on a model to see what adjustments to the design need to be made.

一旦确定了设计款式和面料，就会用比较便宜的材料做出初样（第一次的样衣），并在模特儿身上试穿，以确认需要在设计上做哪些调整。

3. CAD allows designers to view designs of clothing on virtual models and in various colors and shapes, thus saving time by requiring fewer adjustments of prototypes and samples later.

计算机辅助设计能够使设计师查看其设计的衣服穿在电脑虚拟模特上的效果，并可以变换不同的颜色和形状，以此减少对初样和之后样衣的调整，从而节约了时间。

4. In large design firms, fashion designers often are the lead designers who are responsible for creating the designs, choosing the colors and fabrics, and overseeing technical designers who turn the designs into a final product.

在一些大型设计公司，时装设计师通常是首席设计师，负责设计款式、挑选颜色和面料以及监督技术设计师，后者负责将设计稿转变为实际的成品。

5. Costume designers, however, perform extensive research on the styles worn during the period in which the performance takes place, or they work with directors to select and create appropriate

attire.

然而，戏服设计师需要深入研究表演过程中人们的着装风格，或者与导演一起合作，挑选和设计合适的服装。

6. Freelance designers tend to work longer hours and in smaller, more congested environments, and are under pressure to please clients and to find new ones in order to maintain a steady income.

自由设计师工作时间更长，工作环境更狭小、更拥挤。同时，为了维持一份稳定的收入，他们还有取悦老客户、吸引新客户方面的压力。

7. Designers also can gain exposure to potential employers by entering their designs in student or amateur contests.

设计师们还可以通过参加学生或业余设计竞赛，提交他们的设计作品，以提高曝光度和吸引潜在的雇主。

8. Designers must have a strong sense of the aesthetic—an eye for color and detail, a sense of balance and proportion, and an appreciation for beauty.

设计师必须具备强烈的审美意识——对颜色和细节的独到眼光、出色的平衡感和比例感和对美的鉴赏能力。

9. Beginning fashion designers usually start out as pattern makers or sketching assistants for more experienced designers before advancing to higher level positions.

新手设计师起初会为更有经验的设计师做样板师或草绘助手，然后再逐步晋升到更高的职位。

Post-Reading Exercises

- **Reading Comprehension**

Directions: *Read the passage and answer the following questions.*

1) Why do some fashion designers rely on trend reports?

2) What is the normal design process?

3) What are the advantages of computer-aided design (CAD)?

4) What is the creation of high-fashion designers?

5) What is the difference between the work of costume designers and other fashion designers?

6) What do those freelance designers tend to do?

7) For aspiring fashion designers, how do they learn necessary skills and gain valuable experience?
_____

8) Fashion designers must have a strong sense of the aesthetic. What is it?
_____

9) For fashion designers, what is the decisive factor in getting a job?
_____

10) What positions may experienced designers advance to?
_____

- **Vocabulary**

Directions: *Complete the following sentences with the proper forms of the words given in the box.*

| internship | prototype | sketch | amateur |
| congested | costume | preliminary | accessory |

1) Many Beijing residents are now buying electric bicycles to avoid wasting time on _____ streets.

2) We offer products such as handbags, shoes, watches and _____ at the lowest prices.

3) This position is offered initially as a 3 month _____ with the possibility of a full-time position at the end.

4) The most popular Halloween _____ for kids this year: girls, princesses are big, and for boys, pirates are big.

5) Da Vinci's mechanical drawings included _____ designs for such modern contraptions as helicopters and tanks.

6) The talks are at a _____ stage and there is no certainty they will lead to a settlement.

7) The artist _____ the pattern in charcoal (炭笔) on the cloth.

8) All these sports are American inventions and have developed wide support in schools, _____ and professional leagues.

- **Translation**

Directions: *Translate the following sentences into English, using the expressions in the brackets.*

1）我是一个自由职业艺术家、设计师，偶尔做美术教员。（freelance）
_____

2）他们同意成立一个监管机构，对世界金融系统的问题予以警示。（supervisory）

3）这件服装的设计迎合了大众市场需求，在商业上大获成功。（mass market, cater to）

4）我们正在考虑和另一家供货商签约。（supplier）

5）改变织物的质地会大大改变织物的悬垂性。（fabric, drape）

Passage 2   Becoming a Fashion Designer

1   You know you're destined to be a fashion designer if you: a) spent most of your childhood making clothes for your Barbie dolls instead of playing with your friends; b) read fashion magazines instead of your school books; c) ran a boutique out of your basement at age 10. In other words: if you want to be the next Yves Saint Laurent, it helps to be completely and utterly obsessed with fashion.

2   However, there are many aspects of the profession. Working as a fashion designer can just as well mean supervising a design team at a sportswear company as producing a label under your own name. Although the former career may not seem as glamorous as the latter, it certainly will make your life less stressful. To create your own label takes a lot of time, dedication and hard work. Not to mention living just above the poverty line for several years.

Choosing a strategy

3   There are as many different ways to embark upon a fashion career as there are styles of design. Ralph Lauren's Polo empire was founded on a small tie collection that he sold to Bloomingdales. Helmut Lang decided to open his own clothing store when he couldn't find a T-shirt that he liked. Michael Kors built up a network of customers by selling clothes in a trendy NYC boutique. However, most people find that the best foundation for a design career is to get a fine arts degree in fashion at a prestigious school. Besides teaching you the craft, a good school will also add credibility to your resumé. "We live in a brandname society, and having the name of a good school behind you really does help," says Carol Mongo, Director of the Fashion Department at Parsons School of Design in Paris.

Applying to a school

4   There are a lot of colleges that have fashion programs, but only a handful have the kind of reputation that can really push your career. It's hard to enter these schools as competition is high, and they tend to be very selective. You apply by sending a portfolio of drawings of your designs. "We can't teach you how to be creative-you have to bring your creativity to us and let us lead you on your way," says Carol Mongo. She recommends students get some sewing experience before they apply. Drawing is also an important skill for a designer-it is the way you communicate your ideas. In order to build an impressive portfolio it's a good idea to have some experience in sketching; taking art classes will help you understand form

and proportion. But you don't have to be an expert drawer to get accepted to a school. "The most important quality that we look for in our students is that they are truly passionate and exuberant about fashion," says Mongo. "If you have wonderful ideas but can't draw, there are always ways to get around it. You could for example put your designs on a mannequin and take pictures of it."

What school will do for you

5 Most fashion programs are three to four years long. During that time you will take fine arts classes and study drawing, color composition and form. You will also learn pattern making, draping and cutting techniques. One of the most important advantages of design schools is that they work really closely with the industry. Parsons, for example, have "designer critic projects" where successful designers like Donna Karan and Michael Kors work directly with the graduating students. Ambitious students also have the chance to win prestigious awards and grants, which bring them a lot of attention as well as financial support. One very important event is the fashion show at the end of the last semester, when graduating students show their collections. A lot of important people from the fashion industry attend these shows to scout new talent. It's also an opportunity to be really outrageous and get noticed by the media. Hussein Chalayan, for example, became instantly infamous when he showed rotting clothes that he had buried in his backyard for his graduation show at Saint Martins.

Alternative routes

6 "Let's be realistic," says Carol Mongo at Parsons, "School's not for everyone. If you're just looking to get a job in the fashion industry – not a career as a designer – you probably don't need to go school." If you want to work as a seamstress or a patternmaker, the best thing is probably to apply for an internship at a fashion house and work your way up. However, there are many examples of famous designers who started out as interns with no formal training. For example, Dior's brightest new star, men's wear designer Hedi Slimane, had a degree in journalism when he started working with men's wear designer José Levy. Balenciaga's Nicolas Ghesquière is another example of a brilliantly successful designer who learned the jobs hands-on, as an assistant at Jean-Paul Gaultier. Usually, you apply for an internship by sending a portfolio to a fashion house you're interested in. But it's a good idea to call them up beforehand to see exactly what they need. It's also important to note that competition is fierce, and unless you have personal connections, it's very difficult to get an internship without an education.

7 There are also designers, like Luella Bartley, who started their own business after working as stylists for several years, thus building an industry network as well as a good marketing sense.

Understanding the business

8   Unfortunately, it's not enough for a designer to be creative; you also have to have some business sense. As fashion gets more and more corporate driven, it's important to be aware of the business climate and understand the mechanics behind it. By religiously reading trade papers like "Women's Wear Daily" you will get a lot of valuable information. If you want to run your own company, you need to be extremely organized and learn at least the basics of economics. A lot of fashion schools are currently increasing business classes in their curriculum. "Our students have to be smart enough to know how to negotiate a contract, or to pick a business partner," says Carol Mongo. It's perhaps telling that many of the designers that are really successful today, like Calvin Klein or Tom Ford, are involved in every aspect of the business-from licensing strategies to ad campaigns, to actually designing the clothes.

(1067 words)

Proper Names

Bloomingdales	美国布鲁明戴尔百货商店
Calvin Klein	卡尔文·克莱恩，美国著名时装设计师，CK品牌创始人
Carol Mongo	卡罗尔·蒙戈
Donna Karan	唐纳·卡兰，美国著名时装设计师，DKNY服装品牌的创始人
Hedi Slimane	艾迪·斯理曼，法国著名时装设师
Helmut Lang	海尔姆特·朗，著名服装设计师
Hussein Chalayan	侯塞因·卡拉扬，英国时装奇才，先锋艺术（Avant Garde）接班人
Jean-Paul Gaultier	让·保罗·高提耶，法国著名设计师，风格大胆诡异
Jose Levy	荷西·李维，法国著名时装设计师
Luella Bartley	卢埃拉·巴特利，伦敦新锐设计师，曾任英国版 *VOGUE* 杂志服装编辑
Michael Kors	迈克·科尔斯，美国著名服装设计师
Nicolas Ghesquièr	尼古拉·盖斯奇埃尔，巴黎世家（Balenciaga）现任设计师，也是巴黎世家著名的机车包的设计者
NYC (=New York City)	纽约
Ralph Lauren	拉尔夫·劳伦，美国史上最成功设计师一，Polo衫创造者
Saint Martins	（Central Saint Martins College of Art and Design）中央圣马丁艺术与设计学院
Tom Ford	汤姆·福特，美国著名时装设计师，古驰（GUCCI）与圣罗兰（YSL）品牌创意总监

Yves Saint Laurent　　　　　　伊夫·圣·罗朗，世界著名设计大师，法国时尚界的传奇人物，圣罗兰品牌创始人

New Words

beforehand /bɪˈfɔːhænd/ *adv.* 事先，预先
boutique /buːˈtiːk/ *n.* 时装用品小商店；流行女装商店；时装精品店
brand name /ˈbrændneɪm/ *n.* 商标，品牌名称
collection /kəˈlekʃ(ə)n/ *n.* 作品集
craft /krɑːft/ *n.* 手艺，工艺；技巧，技能
credibility /ˌkredəˈbɪləti/ *n.* 可靠性，可信性
curriculum /kəˈrɪkjʊləm/ *n.* 教学大纲，全部课程
destine /ˈdestɪn/ *vt.* 命定，注定；指定
drape /dreɪp/ *vt.* 立体裁剪；*n.* （面料的）悬垂
exuberant /ɪgˈzjuːbərənt/ *adj.* 精力充沛的，热情洋溢的
fierce /fɪəs/ *adj.* （竞争等）激烈的，强烈的；（人或动物）凶猛的，凶残的
form /fɔːm/ *n.* 构造，形态
glamorous /ˈglæmərəs/ *adj.* 迷人的，富有魅力的
grant /grɑːnt/ *n.* 助学金，补助金
handful /ˈhændfʊl/ *n.* 少数人（或事物）；一把（的量）；*adj.* 一把；少数
hands-on *n.* 实际动手经验
infamous /ˈɪnfəməs/ *adj.* 臭名昭著的，声名狼藉的
intern /ˈɪntɜːn/ *n.* 实习生
journalism /ˈdʒɜːnəˌlɪzm/ *n.* 新闻学
mannequin /ˈmænɪkɪn/ *n.* 人体模型，时装模特
outrageous /aʊtˈreɪdʒəs/ *adj.* 令人吃惊的，出人意料的；无礼的，令人无法容忍的
passionate /ˈpæʃnət/ *adj.* 充满激情的
prestigious /preˈstɪdʒəs/ *adj.* 受尊敬的，有名望的，有威信的
proportion /prəˈpɔːʃn/ *n.* 大小比例
religiously /rəˈlɪdʒəsli/ *adv.* 笃信地，虔诚地
resume /ˈrezjʊmeɪ/ *n.* 简历，履历；/rɪˈzjuːm/ *vt.&vi.* 重新开始，重新获得
rotting /ˈrɒtɪŋ/ *adj.* 腐烂的
scout /skaʊt/ *vt.* 物色（优秀音乐家、艺术家等），寻找
seamstress /ˈsemstrɪs/ *n.* 女裁缝
selective /sɪˈlektɪv/ *adj.* 精心选择的
stylist /ˈstaɪlɪst/ *n.* 造型师

trendy /'trendi/ *adj.* 时髦的，赶时髦的，追随时髦的
utterly /'ʌtəli/ *adv.* 完全地，彻底地，绝对地，十足地

Phrases and Expressions

ad campaign	广告宣传活动
be obsessed with	沉迷于
color composition	色彩构成
embark upon	开始，从事，着手
fashion show	时装表演
licensing strategy	授权策略，许可策略
not to mention	更不用说
personal connection	人脉关系
poverty line	贫困线，贫穷线

Key Sentences

1. Working as a fashion designer can just as well mean supervising a design team at a sportswear company as producing a label under your own name.

从事时装设计师这份职业，你可以生产以自己名字命名的品牌，还可以在一个体育用品公司监管一个设计团队。

2. Besides teaching you the craft, a good school will also add credibility to your resume.

一个好的学校，除了教会你技艺外，还会为你的简历增添可信度。

3. There are a lot of colleges that have fashion programs, but only a handful have the kind of reputation that can really push your career.

许多大学都设有服装专业，但只有为数不多的学校拥有良好的口碑，并能真正推动你的事业发展。

4. Ambitious students also have the chance to win prestigious awards and grants, which bring them a lot of attention as well as financial support.

有抱负的学生还有机会赢得一些颇有声望的奖项和奖金，帮助他们吸引大量的关注，同时获得经济上的支持。

5. It's also important to note that competition is fierce, and unless you have personal connections, it's very difficult to get an internship without an education.

此外，有一点必须强调，竞争是非常激烈的，除非你拥有一定的人脉关系，否则要想在没有任何专业背景的前提下获得一次实习机会是非常困难的。

6. As fashion gets more and more corporate driven, it's important to be aware of the business

climate and understanding the mechanics behind it.

随着时装变得越来越企业化，设计师对商业环境的意识和对其背后隐藏的商业运行机制的理解就变得重要起来。

Notes

1. About the author: Fashion Designer Omar Ejaz, the owner of The Heer boutique, has previously exhibited in Lahore, Karachi, New Delhi, Hong Kong, London, Glasgow etc., where his creations were highly appreciated. Omer works with a variety of fabrics, and what truly sets them apart is the innovative way he combines two or more textures in a single garment.

2. Dior（迪奥）: is a French company which owns the high-fashion apparels and accessories producer and retailer Christian Dior Couture, as well as holding 42% of LVMH (Moët Hennessy • Louis Vuitton), the world's largest luxury goods firm.

3. Balenciaga（巴黎世家）: is a fashion house founded by Cristóbal Balenciaga, a Spanish fashion designer. The brand is most famous for its line of motorcycle-inspired handbags, especially the famous "Lariat." It is also very well known for creating avant-garde structural pieces, straddling the edge of fashion and forecasting the future of women's ready-to-wear fashion.

4. *Women's Wear Daily*（WWD）（《女装日刊》）: is a fashion-industry trade journal, sometimes called "the bible of fashion". *WWD* delivers information and intelligence on changing trends and breaking news in the fashion, beauty, and retail industries with a readership composed largely of retailers, designers, manufacturers, marketers, financiers, media executives, advertising agencies, socialites, and trend makers.

Post-Reading Exercises

• **Reading Comprehension**

Directions: *Decide whether the following statements are true or false. Write "T" for true and "F" for false.*

_____1) Working as a fashion designer and producing a label under your own name will definitely make your life more stressful.

_____2) Ralph Lauren opened his own clothing store when he couldn't find a T-shirt.

_____3) If you want to be accepted by a prestigious school, you must be an expert drawer.

_____4) One of the most important advantages of a design school is that it works closely with the industry.

_____5) School is the only way for you to become a successful fashion designer.

_____6) To become a successful fashion designer, you also need to have some business sense.

- **Vocabulary**

Directions: *Complete the following sentences with the proper forms of the words given in the box.*

| boutique | fierce | glamorous | grant | handful |
| infamous | passionate | resume | selective | trendy |

1) I've had a look at the dresses in the new_____, but they're nothing to write home about.

2) Today, the Summer Palace is attracting tourists worldwide with its _____ ancient garden and impressive vigor.

3) A _____ of new books are stirring this debate.

4) Their dog was so _____ that no one dared come near it.

5) The person who wins gets a _____ to study and paint anywhere in England.

6) I have sent a cable describing the wickedness of _____ organization.

7) I need to prepare my _____ for a new job position.

8) True happiness comes when you do what you are most _____ about.

9) The slender cut of this _____ style make the legs appear longer-an effect which can be emphasized with high heels.

10) She is _____ about the clothes she buys.

- **Translation**

Directions: *Translate the following sentences into English, using the expressions in the brackets.*

1）时装设计师可以创建以自己名字命名的品牌。（under one's own name）

2）有些人认为，对于一个国家而言，有一大部分青年人读大学是比较好的。(a large proportion)

3）当我们从事任何工作时，好的开始是很重要的。（embark on）

4）这样导致许多女人吸收了太多的卡路里，更别提大量的垃圾食品了。（not to mention）

5）香奈儿是世界最有名望的时装屋之一。（Chanel, fashion house）

Passage 3 Coco Chanel: The Legend and the Life

1 Coco Chanel's early life and her success as a couturier in Paris equipped her with an enviable little black book, bulging at the seams with the names of the rich and famous. Some were merely clients, but many others were former lovers, quite a few of whom were married. She also had affairs with a number of high-profile men.

2 Two names stand out: the Duke of Westminster Hugh Grosvenor, who she was once tipped to marry, and British prime minister Winston Churchill, connections which may have saved her skin in the aftermath of the Second World War.

3 During the war, Chanel, who was then in her late 50s, had taken a Nazi lover. Hans Gunther von Dincklage, who was 13 years her junior, was reportedly under the direct orders of the Reich Ministry of Propaganda, using a press attachè post in Paris as cover. There was also a theory that he was a double agent, secretly working against the Nazis.

4 In post-Liberation Paris, Chanel was summoned by the French Interior Forces to answer questions. Which of the two, Grosvenor or Churchill, stepped in on her behalf has never been clarified but Coco was soon back at home (in the Ritz Hotel), having been saved the humiliation of having her head shaved and being paraded naked through the streets of Paris along with the other "les collaborations horizontals".

5 When later asked if she had been involved with a German, Chanel replied: "Really, sir, a woman of my age cannot be expected to look at his passport if she has a chance of a lover." The author reflects that "perhaps she was unable to see her German lover without obscuring something of the truth, closing her eyes to his past, as well as his passport; just as she had been apparently blind to previous episodes in her own life".

6 And of these, there were many. Gabrielle Bonheur Chanel was born illegitimate in 1883 in a poorhouse in rural France, and after her mother's death, while she was still a child, her father abandoned her in a convent. Later, details from the nun's habits cropped up in her designs, in particular, collars, cuffs and the use of monochrome black and cream.

7 Coco had a knack for removing the unhappy parts of her earlier life when she had achieved success as a couturier later on. When she was 18 she left the nuns and started working in Moulins as a cabaret singer. She became the mistress of the Etienne Balsan, who introduced her to an English playboy, Boy Capel, who was her lover and muse. She was soon cutting up his polo clothes and turning them into her androgynous

style.

8   With his backing, she was soon on her way as a designer. By 1919, the year he was killed in a car crash, she had her own *maison*. After being a kept woman in the provinces with Etienne Balsan, she had now moved into a rich, royal circle and her world was peopled by exotics like Salvador Dali, Picasso, Cocteau, Diaghilev and another lover, Grand Duke Dmitri, cousin of the Russian tsar. She was back being a mistress again, a role she clearly had no moral problems with as evidenced by her series of relationships.

9   Chanel's clothes mirrored her own lifestyle, replacing corsets with loose trousers and offering a sophisticated liberation, independence, and freedom to women. She replaced ornate evening gowns with her famous little black dress, she popularised costume jewellery, monochrome dressing and her signature collarless, edge-to-edge jacket. She was a huge success as well as a celebrity. But the fallout from the war years effectively brought her career to a halt.

10   Coco closed down her business two hours after war was declared. "This is not the time for fashion," she declared. When she attempted a comeback collection in February 1954, she chose the fifth day of the month-her lucky number-but it didn't do anything to soften the harshness with which she was judged.

11   It was the world of celebrity that turned the tables for Chanel in the US, where she was hailed as the face behind the most famous perfume in the world (Marilyn Monroe famously said the only thing she wore in bed was Chanel No.5). Revenue from the perfume bank rolled the clothing collections and her reputation as a designer recovered.

12   She was, of course, famous for her Chanel suits. The most famous was probably the vivid pink suit Jackie Kennedy wore in Dallas in 1963 on the day President John F. Kennedy was shot beside her in the car. She didn't remove it for a day and a night, forcing the world to see the horror of what had been done. It is now in storage, still caked in the blood of her husband.

13   Picardie acknowledges that in the tales told about Coco-the gossip and speculation and rumours that have spread from newsprint to the internet—the accusation that is repeated most often is that "Chanel was a Nazi collaborator, whose wartime affair with a German officer leaves her reputation blemished, and the legacy of her visionary fashion designs forever stained".

14   However, the author argues that Chanel's "conduct should also be seen in the context of an era of French history marked by a widespread sense of chaos, confusion and uncertainty, as well as terrible tragedy". Chanel admitted at the end: "One shouldn't live alone, it's a mistake. I used to think I had to make my life on my own, but I was wrong."

15   For a woman who revolutionised how women dress and looked to men's wardrobes for many of her innovative ideas, the 20th century icon missed out on the simple pleasures that most women take for granted, a loving, faithful partner and children.

16   As a business woman, she was razor sharp but her private life revealed a very vulnerable

woman who used sleeping tablets and morphine as her last defense against the night.

(997 words)

Proper Names

Boy Capel	"男孩"卡保，英国著名的花花公子
Cocteau	谷克多，法国诗人、小说家、设计家、艺术家
Dallas	达拉斯
Diaghilev	李阿吉列夫，来自俄罗斯芭蕾团
Duke of Westminster Hugh Grosvenor	威斯特敏斯特公爵：休·格罗夫纳
Etienne Balsan	耶田·巴桑
French Interior Forces	法国内务部队
Grand Duke Dmitri	俄罗斯德米特里大公
Hans Gunther von Dincklage	汉斯·君特·凡·丁克拉格男爵
Jackie Kennedy	杰奎琳·肯尼迪，约翰·F.肯尼迪之妻
John F. Kennedy	约翰·F.肯尼迪，美国第35任总统
Marilyn Monroe	玛丽莲·梦露
Moulins	巴黎红磨坊夜总会
Nazi	纳粹
Picasso	毕加索，西班牙著名画家
Reich Ministry of Propaganda	帝国宣传部
Ritz Hotel	丽兹酒店，位于巴黎，全球最奢华的酒店之一
Salvador Dali	萨尔瓦多·达利，西班牙著名超现实主义画家
Winston Churchill	温斯顿·丘吉尔，英国杰出政治家

New Words

abandon /əˈbændən/ vt. 抛弃

accusation /ˌækjuˈzeiʃn/ n. 指责，指控，控告

aftermath /ˈɑːftəmæθ/ n. （战争、事故、不快事情的）后果，创伤

airbrush /ˈeəbrʌʃ/ v. 用喷枪喷；n. 喷枪

androgynous /ænˈdrɔdʒnəs/ adj. 性别特征不明显的，中性的

apologist /əˈpɔlədʒist/ n. 辩解者，辩护者

backing /ˈbækiŋ/ n. 支持，后援

bank /bæŋk/ n. 系列，组，库

blemish /ˈblemiʃ/ vt. 有损……的完美，玷污

bulge /bʌldʒ/ vi. 膨胀，凸出，鼓起
cabaret /'kæbərei/ n. 卡巴莱，有歌舞表演的夜总会；（餐馆、夜总会等处的）歌舞表演
cake /keik/ vt.&vi. （使）结块；（使）胶凝；（厚厚一层干后即变硬的软东西）覆盖
celebrity /sə'lebrəti/ n. 名人，名流
clarify /'klærəfai/ v. 说明，讲清楚；阐明，澄清
chaos /'keiɔs/ n. 混乱，紊乱
collaborator /kə'læbəˌreitə/ n. 通敌者，合作者
collarless /'kɔlələs/ adj. 无领的
convent /'kɔnvənt/ n. 女修道院
corset /'kɔ:sit/ n. 紧身胸衣
couturier /ku:'tjuəriei/ n. 女装设计
cuff /kʌf/ n. 袖口，护腕
enviable /'enviəb(ə)l/ adj. 令人羡慕的
episode /'episəud/ n. （人生的）一段经历；（小说的）片段；（电视剧的）一集
exotic /ig'zɔtik/ n. 外来物，外来的人；adj. 具有异国情调的
fallout /ˌfɔ:'laut/ n. 后果，余波
gossip /'gɔsip/ n. 绯闻
hail /heil/ vt. 赞扬，称颂
halt /hɔ:lt/ n. vt.&vi. 暂停
harshness /'ha:ʃnis/ n. 严厉，苛求
high-profile adj. 引人注目的
humiliation /hju:ˌmili'eiʃn/ n. 羞辱，耻辱
icon /'aikɔn/ n. 偶像，崇拜对象
illegitimate /ˌilə'dʒitəmət/ adv. 私生的，非法的
innovative /'inəveitiv/ adj. 革新的，有改革精神的，引进新观念的
knack /næk/ n. 技能，本领
legacy /'legəsi/ n. 遗留之物，遗产
loose /lu:s/ adj. 宽松的
mansion /'mænʃn/ n. 豪宅，公馆
mirror /'mirə/ vt. 反映
mistress /'mistrəs/ n. 情妇
monochrome /'mɔnəˌkrəum/ adj. 单色的
morphine /'mɔ:fi:n/ n. 吗啡
muse /mju:z/ n. 缪斯；（希神）文艺、美术、音乐等的女神；灵感
naked /'neikid/ adj. 裸体的
nun /nʌn/ n. 修女

obscure /əb'skjuə/ vt. 使……模糊不清，使隐晦，使费解，掩盖；
adj. 不易看清的，暗淡的；费解的，难以理解的
ornate /ɔ:'neit/ adj. 装饰华丽的
people /'pi:pl/ vt. 居住于；使……充满居民；把……挤满人；住满居民
razor /'reizə/ n. 剃刀，剃须刀，刮脸刀
revenue /'revənju:/ n. 收入
revolutionise /ˌrevə'lu:ʃənaiz/ vt. 使彻底变革
revolutionize /ˌrevə'lu:ʃənaiz/ vt. 使彻底变革
rumour /'ru:mə/ n. 谣言
seam /si:m/ n. 接缝处
speculation /ˌspekju'leiʃn/ n. 推测
stained /steind/ adj. 玷污的，褪色的
solidarity /ˌsɔli'dærəti/ n. 团结，齐心协力
summon /'sʌmən/ vt. 传讯（出庭），传唤；召集
tsar /za:/ n. 沙皇
visionary /'viʒənri/ adj. 有远见的
vulnerable /'vʌlnrəbl/ adj. 脆弱的，敏感的；易受伤的，易受责难的
wardrobe /'wɔ:drəub/ n. 衣橱，衣柜

Phrases and Expressions

a double agent	双重间谍
affair with sb.	和某人的恋爱事件，风流韵事
be involved with sb.	与某人关系亲密
be tipped to do sth.	就（某人或某事物）提出意见/建议/（认为是，被猜测是）
crop up	（突然）发生，（意外地）发现
cut up	剪成碎片
equip sb. with sth.	提供给人某物
les collaborations horizontals	（法）通敌者
on one's behalf	代表某人的利益
press attaché post	媒体专员
save one's skin	（俚）使安然无恙；毫发无损
take ... for granted	认为理所当然，想当然
turn the table	扭转形势，转败为胜

Key Sentences

1. Coco Chanel's early life and her success as a couturier in Paris equipped her with an enviable little black book, bulging at the seams with the names of rich and famous.

可可·香奈儿早期的生活，以及在巴黎作为一名女装设计师获得的巨大成功，使她配备了一本令人羡慕的小黑书，里面记满了富人和名人的名字。

2. Which of the two, Grosvenor or Churchill, stepped in on her behalf has never been clarified but Coco was soon back at home (in the Ritz Hotel), having been saved the humiliation of having her head shaved and being paraded naked through the streets of Paris along with the other "les collaborations horizontals".

虽然至今并未澄清，到底是格罗夫纳还是丘吉尔介入并为她说话，香奈儿很快就返回丽兹酒店的家中，避免了遭受被剃头以及和其他"同伙"一起在巴黎的大街上裸体游街示众的羞辱。

3. Later, details from the nun's habits cropped up in her designs, in particular, collars, cuffs and the use of monochrome black and cream.

后来，许多修女服装的细节出现在她的设计中，尤其是领口、袖口和单一黑色和乳白色的运用。

4. Chanel's clothes mirrored her own lifestyle, replacing corsets with loose trousers and offering a sophisticated liberation, independence and freedom to women.

香奈儿的服装是她自己生活方式的真实写照，脱掉了紧身的胸衣，穿上了宽松的裤装。她的服装为女性带来一种糅合了解放、独立和自由的体验。

5. Revenue from the perfume bank rolled the clothing collections and her reputation as a designer recovered.

来自于香水产品的收入带动了其服装产品，最终，她作为设计师的名誉得到恢复。

6. It is now in storage, still caked in the blood of her husband.

这件套装现在仍被保存，上面至今仍留有她丈夫的斑斑血迹。

7. However, the author argues that Chanel's "conduct should also be seen in the context of an era of French history marked by a widespread sense of chaos, confusion and uncertainty, as well as terrible tragedy".

然而，作者却认为，香奈儿的"行为应被放入当时法国大时代背景中去评价。那时期的法国，到处充满混乱、困惑和不确定性，还有可怕的悲剧。"

8. For a woman who revolutionised how women dress and looked to men's wardrobes for many of her innovative ideas, the 20th century icon missed out on the simple pleasures that most women take for granted: a loving, faithful partner and children.

香奈儿，她革命了女人的着装，许多富有创意的设计理念来自于男人的衣柜，但这位20世纪的时尚偶像却没有得到对于大多数妇女来说唾手可得的简单幸福——一位爱她、对她忠诚的伴侣和一群孩子。

9. As a business woman, she was razor sharp but her private life revealed a very vulnerable woman who used sleeping tablets and morphine as her last defense against the night.

作为一名生意人，她一路披荆斩棘，但在私底下，她是一位非常脆弱的女性，需要借助安眠药和吗啡来抵抗黑夜的侵袭。

Notes

1. *Coco Chanel: The Legend and the Life:* The book was written by Justin Picardie and published in 2010. In *Coco Chanel,* Justine Picardie peels away the layers of romance and myth surrounding the legend of Coco Chanel, revealing the true history of the incredible woman who shaped modern fashion and created an empire of haute couture. Picardie's unprecedented research illuminates Chanel's path from little-known seamstress to the aristocracy of style in this stunning look at the fashion icon.

2. Justine Picardie: Justine Picardie is the former features director of *British Vogue*《英国时尚》, and is currently a fashion columnist who writes for several newspapers and magazines, including the *Sunday Telegraph*《星期日电讯报》, *Harpers Bazaar*《时尚芭莎》（英文版）and *Red*《红》. Her bestselling memoir, *IF THE SPIRIT MOVES YOU,* was published by Picador in 2002. She now lives with her husband and two sons in London.

3. Coco Chanel（可可·香奈儿）（1883—1971）was a French fashion designer and founder of the Chanel brand, whose modernist thought, practical design, and pursuit of expensive simplicity made her an important and influential figure in 20th-century fashion. She was the only fashion designer to be named on "Time 100: The Most Important People of the Century".

4. Little black dress（香奈儿小黑裙）is often simply referred to as the "LBD". It is an evening or cocktail dress, cut simply and often quite short. Fashion historians ascribe the origins of the little black dress to the 1920s designs of Coco Chanel. LBD is considered essential to a complete wardrobe by many women and fashion observers, who believe it a "rule of fashion" that every woman should own a simple, elegant black dress that can be dressed up or down depending on the occasion.

5. Chanel No. 5（香奈儿五号香水）was the first fragrance from Parisian couturier Coco Chanel, and has been on sale continuously since its introduction in 1921. It has been described as "the world's most legendary fragrance", and remains the company's most famous perfume. The company estimates that a bottle is sold worldwide every 55 seconds.

Unit 6 Fashion Designers

Post-Reading Exercises

- **Reading Comprehension**

Directions: *Decide whether the following statements are true or false. Write "T" for true and "F" for false.*

_____1) It was Winston Churchill who saved Chanel's skin in the aftermath of the Second World War.

_____2) Chanel, in her late 50s, had taken a Nazi lover, who was 13 years older than she.

_____3) In Chanel's designs, the collars, cuffs, and the use of monochrome black and cream have connected with her early life in a convent.

_____4) At 18, Chanel left the nunnery and started working as a fashion designer.

_____5) In February 1954, Chanel released her comeback collection and regained success.

_____6) The most famous Chanel suit is the vivid pink suit of Jackie Kennedy.

- **Vocabulary**

Directions: *Work in pairs to decide which dictionary entry is the correct meaning for the words given. Put the corresponding number in the space provided.*

abandon /əˌbændən/	*v.* /T/ go away from (a person or thing or place) not intending to return; desert /T/ give up completely (esp. sth. begun) /T/ (~ oneself to sth.) yield completely to (an emotion or impulse)
bulge /bʌldʒ/	*n.* rounded swelling; outward curve (infml) temporary increase in quantity *v.* /vt.&vi. / form a bulge; swell outwards
obscure /əbˌskjuə/	*adj.* not easily or clearly seen or understood; indistinct; hidden not well-known *vt.* make (sth.) obscure(1), hide (sb/sth.)
recover /riˌkʌvə/	*v.* /T/ find again (sth. stolen, lost, etc.); regain possession of sth. /T/ get back the use of (one's faculties, health, etc.) or get back the control of (oneself, one's actions, one's emotions, etc.) /T/ regain (money, time or position)

1) abandon

    A. The match was abandoned because of bad weather.   _____

    B. After his mother died, he abandoned himself to grief.   _____

    C. The car was badly damaged, so they abandoned it.   _____

2) bulge

    A. After the war, there was a bulge in the birth rate.   _____

    B. What is that bulge in your pocket?   _____

C. I can't eat anymore. My stomach's bulging. ＿＿＿＿＿

3) obscure

　　A. The moon was obscured by clouds. ＿＿＿＿＿

　　B. His real motive for the crime remains obscure. ＿＿＿＿＿

　　C. There were many minor and obscure poets in the age of Elizabeth. ＿＿＿＿＿

4) recover

　　A. The police recovered the stolen jewellery. ＿＿＿＿＿

　　B. The team recovered its lead in the second half. ＿＿＿＿＿

　　C. The skater quickly recovered his balance. ＿＿＿＿＿

- **Translation**

Directions: *Translate the following Chinese terms into English.*

1）把衣服剪成碎片 ＿＿＿＿＿＿＿＿＿＿＿＿＿＿＿＿＿＿＿＿＿＿＿＿＿

2）中性的风格 ＿＿＿＿＿＿＿＿＿＿＿＿＿＿＿＿＿＿＿＿＿＿＿＿＿＿＿

3）单色裙装 ＿＿＿＿＿＿＿＿＿＿＿＿＿＿＿＿＿＿＿＿＿＿＿＿＿＿＿＿

4）无领夹克 ＿＿＿＿＿＿＿＿＿＿＿＿＿＿＿＿＿＿＿＿＿＿＿＿＿＿＿＿

5）使她的事业中断 ＿＿＿＿＿＿＿＿＿＿＿＿＿＿＿＿＿＿＿＿＿＿＿＿

6）富有远见的服装设计 ＿＿＿＿＿＿＿＿＿＿＿＿＿＿＿＿＿＿＿＿＿＿

7）男人的衣橱 ＿＿＿＿＿＿＿＿＿＿＿＿＿＿＿＿＿＿＿＿＿＿＿＿＿＿

8）一位忠诚的伴侣 ＿＿＿＿＿＿＿＿＿＿＿＿＿＿＿＿＿＿＿＿＿＿＿＿

Unit 7 Fashion Marketing

PART ONE　Warm-up Activities

Bring Color to Life with a Natural Dye

New Words

dye /daɪ/ n. 染料	vat /væt/ v. 在大桶里染（处理）；n. 大桶
squeeze /skwiːz/ v. 挤出	mordant /'mɔːdnt/ n. 媒染剂，酸洗剂
alum /'æləm/ n. 明矾	tartar /'tɑːtə/ n. 酒石

Directions: *Listen to the passage about a natural way to dye wool and decide whether the following statements are true or false. Write "T" for True and "F" for False in the spaces provided.*

_____1. Most clothing is colored with dyes, and natural dyes can be very expensive.

_____2. The vat method can be used to dye wool with potato skins.

_____3. In the dying process, a mordant helps fix the dye to the material.

_____4. Wood ash is a popular mordant and is often mixed with cream of tartar.

_____5. Keep the wool in the mixture of alum and cream of tartar for forty-five minutes before heating.

_____ 6. Place the wool into the dye and heat the mixture to a boil, and then immediately reduce the heat to eighty-two degrees.

Rise of Domestic Consumer Brands

New Words

franchise /'fræntʃaɪz/ n. 特权，经销权	nimble /'nɪmbl/ adj. 敏捷的，聪明的
catapult /'kætəpʌlt/ vt. 猛投，弹射	forge /fɔːdʒ/ vt. 锻造，伪造，前进

Directions: *Listen to the passage and fill in each blank with the information you get from the recording.*

The rise of domestic Chinese consumer brands to challenge their famous foreign counterparts has become a defining 1._____ in China's industrial development.

Having established the competitiveness of their brands versus foreign 2. _____ , Chinese companies have several advantages. One is that they tend to rely less on franchise stores than their foreign counterparts, giving them greater control over 3. _____ and brand management. Another 4. _____ is that local brands are generally more nimble in expanding into lower-tier cities, a prime source of revenue.

For China's 5. _____ market, for example, it may grow from an estimated US$7.2bn in 2009 to around US$12.4bn in 2012 in value terms. Such growth, if it materializes, would be likely to catapult some Chinese sportswear brands into powerful or even 6. _____ positions in the industry worldwide. And this is true not only for sportswear but also for 7. _____ garments and several other consumer industries as well.

Wang Anbang, president of Susino, one of the world's largest umbrella makers, plans to open several hundred branded stores nationwide over the next year to change his business from merely making umbrellas to selling branded fashion 8. _____ . Susino will aim first at the domestic market, and then seek to build its brand overseas by forging 9. _____ with suitable foreign companies. Mr Wang is also keeping his eye open for suitable foreign acquisitions. "It is no problem for us to make an 10. _____ overseas. China's capital markets have loads of money. It is easy for us to buy," he said. However, the focus would be on learning about overseas markets, fashions and consumer trends rather than charging ahead with acquisitions, Mr Wang said.

PART TWO   Reading Activities

Passage 1   The New Era of Fashion Marketing

by Christopher Ruvo

1   Apparel marketing needs a change.

2   The premise of a product selling itself; of dumping a catalog on a client's desk to pick out one polo in a sea of pages; of ignoring technology and resisting change – we would say those days are coming to an end, except that so many of the most successful distributors have already left that behind.

3   It's time to adapt. So who is best to learn from when it comes to apparel marketing? The fashion industry. Whether it's generating a legion of loyal customers to the brand or turning apparel into a "now" item or hitching its star to the hottest celebrities – no one does it better. We identified three main areas where distributors need to up their game and take cues from those who do it best in the fashion and wearables industries. Welcome to the new era of fashion marketing.

4   Fashion PR, With a brand image and desired audience to target, the PR push begins. Typically, this involves crafting press packages that get sent to decision-makers at print, online and televised media outlets. Sometimes stores are targeted directly, too. Press packages, which should convey the brand's essential story and image, can include samples, photos, videos, examples of other press coverage and, of course, a written news release.

5   To garner coverage and placements, top fashion lines and their PR firms build relationships with key figures in media and entertainment. Stacy Igel, the creator of "Boy Meets Girl", a hip clothing and accessories line for young women, has forged connections that led to her line being covered in publications like *Us Weekly* and *Cosmopolitan*. She has also earned television placements, including a spot on "Gossip Girl". Additionally, Igel has entered into co-promotional partnerships with entertainment figures like pop novelist Caprice Crane. The two teamed up to run a Web-based contest in which the winner received an advance copy of Crane's new novel and a promotional shirt, "Boy Meets Girl", made for the book. "These organic relationships have helped my brand grow," says Igel.

6   Launch parties and fashion shows are other weapons in the fashion PR arsenal. Raygorodskaya,

the president of a New York City public relations firm, says these events must align with a brand's image. She recently orchestrated a show for a high-end clothing line produced by a former professional basketball player. During the show, a screen displayed a door that opened to reveal images of men wearing the clothes – designed for the big and tall – while engaged in high-income lifestyle pursuits. The images played as models strutted on stage. Says Raygorodskaya, "It gave the media a look at what the brand stands for."

7 Brand Marketing, The fashion world knows that brand is king, and appropriately sells the story of a clothing line. When Raygorodskaya works with a client, she often begins by asking, "If your brand was a person, what would they do on the weekend? What magazines would they read?" Having a strong brand personality helps a fashion company define goals and establish a target audience. "The brands that do well are ones people can identify with, but you have to give them something real to connect to," she says.

8 So too have successful distributors realized that by injecting creativity into the mix, they can pair clients with logoed apparel that builds brand identity. At Rightsleeve Marketing, the Toronto-based distributor learns about a brand and the end-users expected to wear its apparel. If given the chance, the distributor designs a unique logo that conveys a client's image. High-tech firms, for example, might get a sleek logo that expresses cutting-edge innovation and precision. A law firm's logo may transmit staidness, reliability. If a logo is established, Rightsleeve focuses on pairing the client with apparel that's tailored to the tastes of intended end-users. This process includes determining proper garment and decoration types, as well as offering, say, shirts in different colors, since end-users desire choice, says Rightsleeve President Mark Graham. Through the distributor's efforts, the client receives logoed apparel that people will actually wear. "The most successful programs we employ put the interest of the end-user first," Graham says.

9 Nonetheless, distributors will likely encounter clients who will not, initially, want to spend the cash necessary to obtain brand-enhancing apparel. Distributors need to stress that companies don't want to damage their brand image by buying substandard apparel that won't last and leaves a bad impression. While respecting budget is paramount, distributors should take a "good, better, best" approach, showing clients what they could receive for a lower-priced shirt and a higher-priced garment. With the higher-end item, there's potential for greater distinction and better ROI.

10 Social Media, "Think. Connect. Listen. Party. Measure." Those are the pillars of Urban Outfitters' social media strategy. The fact that the fashion retailer, which sells trendy clothing and accessories, has such a strategy highlights that online outlets like Facebook, Twitter and YouTube are

now integral components of fashion brands' marketing and retention strategies. Indeed, companies like Urban Outfitters are increasingly using these platforms to build brand identity and forge tighter relationships with consumers. Similarly, distributors are incorporating social media into apparel campaigns for clients and using the media in conjunction with apparel to promote their own companies. "The key to making social media work is to understand what your audience stands for, their values and pain points, and then to connect with them in a relevant way," says Nick.

11 Fashion brands and distributors also use social media to offer sales and giveaways. Motivators, for example, offers coupons to people who "Like" the distributor's Facebook page. Urban Outfitters recently invited Facebook followers to submit travel photos for the chance to win a $100 gift certificate. HauteLook, a flash sale website that sells luxury apparel at discount prices, has offered preview sales exclusively on its Facebook page. The discount site also teamed up with the Diane Von Furstenburg line for a sale of DVF clothing through the HauteLook Facebook site.

12 Growing numbers of distributors have put QR codes on promotional apparel for their own companies and clients. Landmark, for example, used QR codes on a shirt to help raise awareness about a bank's acquisitions and expansion. The codes, when scanned by a smartphone, linked to the bank Web page. "People like to play with their smartphones," says Schlechte. "This takes advantage of that and gets them looking at the bank." Motivators put QR codes on the back of shirts employees wore to an end-user trade show. Once scanned, the codes directed show attendees to a specific section of Motivators' website that contained products that appealed to them. Says Nick: "The ROI was immaculate."

(1125words)

New Words

accessory /ək'sesəri/ *n.* 饰品
acquisition /ˌækwɪ'zɪʃ(ə)n/ *n.* 收购
advanced /əd'vɑːnst/ *adj.* 前进的，先头的，预先的，高级的
align /ə'laɪn/ *v.* 与……保持一致
apparel /ə'pærəl/ *n.* 服装
arsenal /'ɑːs(ə)n(ə)l/ *n.* 军械库，武器库
attendee /ˌəten'diː/ *n.* 参加者，出席者
coupon /'kuːpɒn/ *n.* 赠券，（连在广告上的）预约券，优待券，优惠券
cutting-edge /ˌkʌtɪŋ'edʒ/ *adj.* 前沿的，尖端的

distributor /dɪˈstrɪbjutə/ n. 销售者；批发商
end-user /ˈend-juzə/ n. 最终用户，实际用户
forge /fɔːdʒ/ v. （尤指努力地）生产、制造
garner /ˈgɔː(r)nə/ v. 〈诗〉贮藏，积累
giveaway /ˈgɪvəˌweɪ/ n. （招徕顾客的）赠品
hip /hɪp/ adj. (hipper, hippest)（非正式）时髦的，时尚的
hitch /hɪtʃ/ v. 被挂住，被钩住
immaculate /ɪˈmækjulət/ adj. 毫无瑕疵的，无缺点的
incorporate /ɪnˈkɔːpəreɪt/ v. 结合，合并，收编
integral /ˈɪntɪgrəl/ adj. 完全的；缺一不可的，主要的
legion /ˈliːdʒ(ə)n/ n. 众多，大批，无数
orchestrate /ˈɔːkɪˌstreɪt/ v. 为（管弦乐队）谱写音乐；使和谐地结合起来
outlet /ˈaʊtˌlet/ n. 零售网点，经销点，专卖店；（美）（通常坐落在市郊的）购物中心，廉价商品销售中心
paramount /ˈpærəmaʊnt/ adj. 最高的，至上的，首要的
premise /ˈpremɪs/ n. （理由等的）前提，根据，缘起部分
preview /ˈpriːvjuː/ n. 预观；预映；试映；预演，试演；（展览会的）预展；预习
release /rɪˈliːs/ n. 发表；发售（物）；（影片的）发行上映
retention /rɪˈtenʃ(ə)n/ n. 保留，保持，维持
sleek /sliːk/ adj. 非常时髦的，豪华的；兴旺的
staidness /ˈsteɪdnɪs/ n. 认真；沉着
strut /strʌt/ v. 大摇大摆地走，趾高气扬地走
substandard /ˌsʌbˈstændəd/ adj. 标准以下的
trendy /ˈtrendi/ adj. 时尚的，流行的
wearable /ˈweərəb(ə)l/ n. 衣服，服装

Phrases and Expressions

launch party	开盘仪式，举办会议
PR (public relations)	公共关系
press package	新闻材料
QR code (Quick Response code)	二维码
ROI (Return On Investment)	投资回报率
take one's cue from	学……的样子
team up	合作，协作；结为一对

Key Sentences

1. Whether it's generating a legion of loyal customers to the brand, or turning apparel into a "now" item, or hitching its star to the hottest celebrities – no one does it better.

时尚产业是否能产生一大批忠于品牌的客户，或者是否能把服装转变为一项"时兴"的项目，是否能把它的主角人物发展为最热门的明星——大家都做得不够好。

2. Press packages, which should convey the brand's essential story and image, can include samples, photos, videos, examples of other press coverage and, of course, a written news release.

用于传达品牌核心故事和形象的媒体包装，可以包含样品、图片、视频、其他新闻报道的案例，当然还有书面新闻发布。

3. So too have successful distributors realized that by injecting creativity into the mix, they can pair clients with logoed apparel that builds brand identity.

同样，成功的经销商们也已经意识到，通过注入创意组合能够把客户和构建品牌身份的标识性服装顺利配对。

4. Nonetheless, distributors will likely encounter clients who will not, initially, want to spend the cash necessary to obtain brand-enhancing apparel.

然而，经销商很可能会遇到原本并不打算花钱以提升服装品牌的客户。

5. The fact that the fashion retailer, which sells trendy clothing and accessories, has such a strategy, highlights that online outlets like Facebook, Twitter and YouTube are now integral components of fashion brands' marketing and retention strategies.

销售流行服装和配饰的零售商有这样的一个策略，这一点突显出像 Facebook, Twitter 和 YouTube 等在线商城现在已经是时尚品牌市场营销和保持策略的不可或缺的组成部分。

6. Similarly, distributors are incorporating social media into apparel campaigns for clients and using the media in conjunction with apparel to promote their own companies.

同样的，经销商们正在结合使用社会媒体来争取客户，同时协同服装业使用各种媒体来促进各自的公司发展。

Notes

1. *Us Weekly*: is a celebrity gossip magazine. It was founded in 1977 by The New York Times Company, which sold it in 1980. Then it was acquired by Wenner Media in 1986. The publication covers topics ranging from celebrity relationships to the latest trends in fashion, beauty, and entertainment.

2. *Cosmopolitan*: is an international magazine for women. It was first published in 1886 in the United States as a family magazine, was later transformed into a literary magazines and eventually became a women's magazine in the late 1960s. Also known as *Cosmo*, its content as of

2011 included articles on relationships and sex, health, careers, self-improvement, celebrities, as well as fashion and beauty. Published by Hearst Magazines, *Cosmopolitan* has 63 international editions, is printed in 32 languages and is distributed in more than 100 countries.

3. *Gossip Girl*: is an American teen drama television series based on the book series of the same name written by Cecily von Ziegesar. The series was created by Josh Schwartz and Stephanie Savage, and premiered on The CW on September 19, 2007. Narrated by the omniscient yet unseen blogger "Gossip Girl", the series revolves around the lives of privileged young adults on Manhattan's Upper East Side in New York City.

4. *Caprice Crane*: is an American novelist, screenwriter and television writer/producer. She was born in Los Angeles, California and graduated from New York University (NYU) Tisch School of the Arts Film School. Her novels include Stupid & Contagious (2006), Forget About It (2007), and Family Affair (2009). Her newest novel, With a Little Luck, was published on July 26, 2011 by Bantam Books.

5. This article is adapted from *Wearables* (September 2011).

Post-Reading Exercises

- **Reading Comprehension**

Directions: *Read the passage and answer the following questions.*
1) Why does the writer say "It is time to adapt"?
2) What should be included in press packages in the fashion field?
3) What are the weapons used by fashion PR firms to ensure their clients become known to the public?
4) Why is brand-building vital to the success of a clothing line?
5) What kind of logo is appropriate to a law firm?
6) What should distributors do if their clients lack awareness of building brand identity?
7) What is vital when using the strategy of social media to build brand identity?
8) Can social media help promote product sales?

- **Vocabulary**

Directions: *Complete the following sentences with the proper forms of the words given in the box.*

| acquisition | advance | coupon | giveaway | integral |
| legion | premise | preview | release | futuristic |

1) We're kicking 2012 off right by launching a huge _____ ! Enter our website to win an Apple iPad2!

2) Beijing residents aged 80 or above can receive a monthly _____ from the government worth 100 yuan ($15.6).

3) The cafeteria features sleek chairs and _____ décor.

4) A(n) _____ payment for his surgery in the amount of 7,680 euros was made by his parents through a bank transfer.

5) His retirement from the NBA disappoints his teammates and his growing _____ of fans.

6) The company was founded in 1954 and expanded through a series of _____ topped by a $12.6 billion deal for the Rouse Company in 2004.

7) Filial piety is the foundation of all moralities and the _____ of many ethical concepts in Chinese traditional culture.

8) Technology is a/an _____ part of all people' lives.

9) A pocket watch found in the Titanic wreckage was among a sampling of artifacts on _____ on Thursday in New York.

10) WikiLeaks became the focus of a global debate over its role in the _____ of the secrets about the wars in Iraq and Afghanistan.

- **Translation**

Directions: *Translate the following sentences into English, using the expressions in the brackets.*

1）如果你向她学习的话，总有一天你会成为博学之士。(take the cue from)
_____

2）公司已决定与三星电子合作研发新型智能手机。(team up)
_____

3）每个国家必须找到适合自己独特经济环境的发展策略。(be tailored to)
_____

4）在这件事上，学校的利益与家长的利益并不总是一致。(align with)
_____

5）我们应该将这些有经验的管理人员的建议融入最新的设计之中。(incorporate ... into)
_____

Passage 2 Online Shopping: Fashion at Your Fingertips

Online shopping is booming at the expense of high street stores...
by Jessica Fellowes

1 A few years from now, when the banks are wholly owned by the government, our houses are worth less than our cars and we have to travel by rickshaw, we'll be telling our grandchildren about the days when we used to go to "shops" to buy clothes. How will we explain the phenomenon of whole buildings being used not to house thousands of redundant bankers and estate agents, but to store rail after rail of clothes.

2 Too futuristic? Perhaps. But if the shop isn't quite dead, 2009 will certainly be remembered as the year when it started losing its fight against the internet. IMRG, the online retail industry body, reported a 34 per cent increase in online shopping last year. Industry insiders are predicting that by 2016, the online fashion business will account for 13 per cent of the fashion market and be worth some £6 billion.

3 Dedicated fashion websites such as asos.com, which recorded a 118 per cent year-on-year increase in last December's sales figures, offer cut-price labels and celebrity style. They have become a modern-day phenomenon. "George" at Asda, for example, which launched online last year, already gets more than a quarter of a million customers a week. "Jigsaw", which launched its website last November, also reports healthy figures.

4 Not surprisingly, the big boys have decided they want a piece of the action. This week "John Lewis", which reported traffic on Christmas Day 12 times what it was a year ago, has launched a "Basic Deluxe Collection", which they say yields a higher return on "cost per wear". A sumptuous "Boyfriend" Cardigan in silk/angora (£83) is sure to be on the backs of many a yummy mummy this year.

5 Retail giant "Tesco" has also declared plans to start selling its hugely successful clothing ranges "Florence", "Fred" and "Cherokee" online this autumn. It may seem a long time to wait, but

you can already browse a limited selection at clothing at tesco.com. We can look forward to picking up strong looks such as the current F+F kimono flower-print dresses, (£15) , a striking purple ruffle-bib blouse with coral pencil skirt (£10 and £8), and the double-breasted, big-buttoned navy jacket with white, wide-legged trousers (£15 and £15) .

6   There are plenty of bargains to be found on other sites. "Asos" stock 600 brands, with 500 new items a week. This week our favourites are their Hooded Woven Dress (£27.50), the Woven Harem Pant (£24.47) and the Rik Rak Dress in the style of Alexa Chung (£44.50).

7   "Topshop" is hot to trot, with 300 new items added every week. Madly on trend is the flowery Photographic Corset Dress (£45) and silk All In One Playsuit (£38). New Look's website has a bang-on trend high-waisted Pencil Skirt (£18) and - new in - Boyfriend Fit Jeans (£15).

8   "George at Asda" has a smaller range but is just as quick off the mark. A Lipstick Print Dress with a large bow on the waistband is £18, and the double-breasted military Crop Jacket is a credit-crunch-friendly £12. I've known fish fingers to cost more.

9   The move online seems unavoidable, but a few retailers are still resisting it. So far, "Primark", which says it has "no plans for an online platform", seems unconcerned, as does H&M, which has websites in the Nordic countries but no plans to launch one for UK customers this year. "The focus is on the stores - we are planning 225 new stores worldwide," a spokesperson says.

10   Will delaying mean they miss the fashion boat? "Gap", which sells online in the US, is rumoured to be launching in the UK soon but "they risk losing out to first-mover advantage," says David Smith of IMRG. "Online shoppers have their favourite websites just as they have their favourite High Street shops."

11   "I'd be looking to online shopping for my growth," says Mary Portas, our resident Queen of Shops. But what of her mantra? Do the best retailers provide a total experience? "Online shopping is about accessibility and shopping when you want, how you want. Shopping online is the perfect solution. Middle and lower ends of the retail market will lose out to the internet."

12   But isn't shopping for clothes about top designers and about feeling the quality of the fabric? Not when you can get such on-trend bargains, suggests Siobhan Mallen, associate fashion editor of *Grazia*, the fashion weekly. They often feature supermarket clothes on their pages. "Dressing stylishly is all about the mix," says Mallen.

13   While online shopping scores on convenience, catwalk videos, zoom photography that's

shot from every angle and browsing at 3:00 am when you can't sleep, Mallen also believes online clothes sites are riding high because of "the guilt factor" – it's not the time to be seen with shopping bags from Bond Street. "It looks like showing off," she says. So no more jealous looks on the bus. No spouse asking what's in the bag.

14 Furtive shopping: it's the future.

(854 words)

New Words

angora /æŋ'gɔ:rə/ *n.* 安哥拉兔毛（或山羊毛）线（或织物）
bang-on /ˌbæŋ'ɔn/ *adj.* 极精确的；非常有效的
bib /bib/ *n.* （小儿）围涎，围嘴；围腰的上部
blouse /blauz/ *n.* 女衬衫
cardigan /'kɔ:(r)digən/ *n.* （开襟）羊毛衫，羊毛背心，开襟绒线衫
corset /'kɔ:(r)sit/ *n.* 紧身胸衣
crunch /krʌntʃ/ *n.* （经济等）紧缩状态
deluxe /də'lʌks/ *adj.* 豪华的，奢侈的，高级的
fabric /'fæbrik/ *n.* 编织品，织物
furtive /'fə:tiv/ *adj.* （人）偷偷摸摸的，鬼鬼祟祟的；（行动）秘密的
futuristic /ˌfju:tʃə'ristik/ *adj.* 未来的，未来主义的
kimono /ki'məunəu/ *n.* 和服
mantra /'mæntrə/ *n.* 颂歌，圣歌；咒语
Nordic /'nɔ:dik/ *adj.* 北欧人的
playsuit /'pleisju:t/ *n.* （妇女，儿童）运动衫，运动裤
redundant /ri'dʌndənt/ *adj.* 过多的，冗余的，冗长的（文章等）；累赘的；丰富的（食物等）
rickshaw /'rikʃɔ:/ *n.* 人力车，黄包车
ruffle /'rʌf(ə)l/ *n.* 褶边，褶边状物
spouse /spaus/ *n.* 配偶，夫，妻；夫妇
sumptuous /'sʌmptʃuəs/ *adj.* 奢侈的，豪华的；高价的
trot /trɔt/ *v.* （人）匆忙地走，快步走
waistband /'weis(t)ˌbænd/ *n.* 腰带，裤带
zoom /zu:m/ *n.* （摄影）可变焦距镜头

Phrases and Expressions

be quick /slow off the mark　　　　　　　反应迅速 / 迟钝

double-breasted	（外衣等）双排扣的，对襟的
harem pant	灯笼裤，哈伦裤
hooded woven dress	连帽针织裙
lose out to	输给；被……取代
military crop jacket	军装款短夹克
pencil skirt	铅笔裙，紧身窄裙
wide-legged pants	宽腿裤
yummy mummy	（非正式）漂亮妈妈；辣妈

Key Sentences

1. How will we explain the phenomenon of whole buildings being used not to house thousands of redundant bankers and estate agents, but to store rail after rail of clothes?

整幢大楼不再是很多的银行家和房产经纪人办公的场所，而成了一家又一家的服装存储仓库，我们该怎么解释这种现象呢？

2. A sumptuous Boyfriend Cardigan in silk/angora (£83) is sure to be on the backs of many a yummy mummy this year.

今年在很多漂亮的妈咪身上都可以看到真丝/安哥拉毛制的女式针织衫（价格为83英镑），非常华丽。

3. We can look forward to picking up strong looks such as the current F+F kimono flower-print dresses (£15), a striking purple ruffle-bib blouse with coral pencil skirt (£10 and £8), and the double-breasted, big-buttoned navy jacket with white wide-legged trousers (£15 and £15) and all above.

我们可以期待看到视觉感强劲的搭配，譬如，目前的F+F和服式的印花连衣裙（15英镑），亮紫色荷叶边衬衫配珊瑚色铅笔裙（10英镑和8英镑），双排大纽扣的深蓝色夹克配白色阔腿裤（15英镑和15英镑）以及上面提到的全部。

4. So far, Primark, which says it has "no plans for an online platform", seems unconcerned, as does H&M, which has websites in the Nordic countries but no plans to launch one for UK customers this year.

目前为止，Primark这一品牌看起来并不关注线上销售，也没有任何计划发展在线平台。H&M也是如此，在北欧有一些销售网站，但是今年针对英国的客户没有任何网站发展计划。

5. But isn't shopping for clothes about top designers and about feeling the quality of the fabric?

但是难道购买衣服不是为了关注高端的设计师们以及追寻织物的品质吗？

6. While online shopping scores on convenience, catwalk videos, zoom photography that's shot from every angle and browsing at 3:00 am when you can't sleep, Mallen also believes online clothes sites are riding high because of "the guilt factor" – it's not the time to be seen with shopping bags from Bond Street.

网店得益于便利的特性，凌晨三点你睡不着时可以观看时装表演的视频和不同角度的变焦拍摄，Mallen 还认为网店的优势正在于"这愧疚的因素"——不该是被看到拎着邦德街的购物袋的时间。

Notes

1. ZMRG (Interactive Media in Retail Group) is the UK's industry association for e-retail, a membership community set up to develop and share the latest best practice advice to enable retailers to succeed in the world's most competitive online market. The IMRG site now features a fully interactive platform offering a wide range of insights into the performance of the global e-commerce market. The IMR Smart Knowledge Base enables users to access a wide range of datasets covering all the key areas that make up the online experience, both nationally and internationally.

2. Alexa Chung (born 5 November 1983) is a British television presenter, model and contributing editor at British Vogue. Chung is a Muse to many fashion designers because of her distinctive personal style. She frequently appears on best dressed lists, is a regular cover girl for *Vogue*, *Elle* and *Harper's Bazaar*, and is often seen in the front row at fashion shows.

3. Asda Stores Ltd. is a British supermarket chain which retails food, clothing, general merchandise, toys and financial services. Its head office is at Asda House in Leeds, West Yorkshire. Asda became a subsidiary of the American retail giant Walmart, the world's largest retailer, in 1999, and is the UK's second largest chain by market share after Tesco. In December 2010, Asda's share of the UK grocery market stood at 16.5%.

4. Tesco plc is a global grocery and general merchandise retailer headquartered in Cheshunt, United Kingdom. Founded in 1919, it is the third-largest retailer in the world measured by revenues (after Wal-Mart and Carrefour) and the second-largest measured by profits (after Wal-Mart). It has stores in 14 countries across Asia, Europe, and North America, and is the grocery market leader in the UK, Malaysia, the Republic of Ireland, and Thailand.

5. Bond Street is a major shopping street in the West End of London that runs north-south through Mayfair between Oxford Street and Piccadilly. It has been a fashionable shopping street since the 18th century and is currently the home of many high price fashion shops. The southern section is known as Old Bond Street, and the northern section, which is rather more than half the total length, is known as New Bond Street. This distinction,

however, is not generally made in everyday usage. It is one of the most expensive strips of real estate in the world.

Post-Reading Exercises

- **Reading Comprehension**

Directions: *Read the passage and answer the following questions.*

1) What does the future look like according to the writer?
2) Why does the writer think the shop will die in the future?
3) What does the writer mean by saying "the big boys have decided they want a piece of the action"?
4) How many brands can be browsed in asos.com now?
5) Will delaying online business mean the company misses the fashion boat?
6) What are the merits of online shoping as mentioned by the writer?
7) What is the most important factor in online shopping? Designer or the quality of the fabric?

- **Vocabulary**

Directions: *Complete the following sentences with the proper forms of the words given in the box.*

| crunch | deluxe | dedicate | fabric |
| futuristic | redundant | furtive | spouse |

1) The National Museum of American History plans to _____ a major collection devoted to hip-hop music.
2) The auction house raised $170,000 through the sale of a _____ limited-edition watch.
3) Investment in industries with overcapacity, high pollution, and energy intensity, and _____ construction must be curbed.
4) The _____ CCTV tower – two towers that meet in an arch was designed by German architect Ole Scheeren.
5) The store sells merchandise made of real glass, stainless steel, ceramics and _____ .
6) Her _____ glance at every passer-by signals her constant vigilance.
7) The economic downturn results in a credit _____ that will further hurt the struggling economy.
8) While men attached more importance to appearance in choosing a _____ , women preferred men who were successful in their careers.

- **Translation**

Directions: *Translate the following sentences from Chinese to English, using the expressions in the brackets.*

1）行动敏捷些，你才能在比赛中取得好成绩。(be quick off the mark)

_____

2）金融、电信和商业服务仅占印度服务业产出的28%。(account for)

_____

3）尽管电视越来越普及，但我认为电影不会被电视所取代。(lose out to)

_____

4）他期望在退休以后能与他的家人多聚聚。(look forward to)

_____

5）高产出率往往是以牺牲产品的质量为代价的。(at the expense of)

_____

Passage 3　Zara and H&M: Fast Fashion on Demand

1　When was it that high-fashion became everyday wear? I feel like it happened so fast. Suddenly everybody considered themselves fashion oracles with Jimmy Choo shoes and a bag to match. The people you would least expect had a fashion blog, took pictures of themselves in "Today's Outfit," and had opinions on coming trends, underground designer wear and new fashion shoots. Elle was the new *Times* and wedges were no longer a mathematical term. But obviously  price is a factor in this. Everybody cannot afford a pair of Jimmy Choos. Well, hang on a minute… until H&M and Zara. That's what happened, and that's when everything else followed. They made it possible. They brought the latest catwalk trends to the masses without the accompanying price tag. H&M with their designer collaborations and Zara with their incredible speed participate in ripping off new designer styles.

2　So how did they accomplish this? Well, it is not as simple as copying what celebrities just wore and thinking consumers will be satisfied with that. One reason, especially in Zara's case, is the perceived exclusivity. Zara achieves this through always having new pieces of clothing in limited supply. That there are only a few items on display, even though the average size of a Zara store is 1,000 square meters, helps. And that Zara removes unsold clothes after just a few weeks also contributes towards the exclusivity. They have successfully created a "now or never" customer mentality.

3　Individual Zara stores can also choose autonomously what to order for their specific store, which makes each store special and separate from another. Customers are then encouraged to go to several Zara stores in one single shopping round. But the reason Zara can get those shoes that look exactly like the Balmains out so fast to their stores is their centralized production center in Spain. That market specialists, procurement personnel, and sales people are all sitting next to each other in Spanish La Coruña allows the process - between sketch, through design, to send-off - to be instant. Zara can take a design from the drawing board to stores in only two weeks. This enables them to capture new catwalk trends very

quickly. Zara's 200 designers, free of the common pretensions, sit in the midst of the action. Zara also lets their retailers change orders to a higher extent than competitors. This flexibility saves them overproduction, and using cheaper materials permits the lower price.

4 All this combined with Zara's closely surveyed distribution network, enables them to introduce new fashion products every week, which has created excitement in addition to the exclusivity. They collect data from their 4,430 stores in 73 countries to detect trends and be able to respond with exact precision to consumer preferences. This has led to Zara not needing much advertising—another cost saver in addition to low inventories. Since every day is a surprise and you never know what you will find at Zara (that mysteriously complies exactly with your taste), customers are repeatedly coming back to see what's new.

5 H&M has managed to achieve this same sensation but through a slightly different strategy. About one quarter of their inventory is made by H&M themselves, in comparison to Zara whose whole operation is. H&M keeps lots of clothing inventory originating from cheap Asian manufacturers. So H&M is maybe not that efficient and does not deserve to be held up with Zara in the fast fashion category, or so you might think. How come they have 2,000 stores in 37 countries and employ 76,000 people? The Swedish retailer has paired up with high-fashion designers to create limited collections, creating hysteria among customers, media and uncountable fashion-bloggers, selling out instantly. These designer collaborations with Stella McCartney, Karl Lagerfeld, Viktor & Rolf, Roberto Cavalli, Comme des Garcons, Matthew Williamson, Jimmy Choo and Sonia Rykiel,

have given H&M legitimacy as a new-thinking fashion company and even given them credibility with alternative fashion movements, especially in the U.S. The fact that only selected stores carry the collections creates a "have to have it" feeling similar to what Zara achieves.

6 If Zara is the most efficient, H&M is certainly the pioneer. They were first with using young talented designers to make cheap yet high-fashion items at a fast pace. Although it might not seem like it because of the careful Swedish exterior, H&M are constantly experimenting. With their guest designers for example, they launched into swimwear, menswear, new garments and fabrics that these designer had never worked with before. Jimmy Choo had never done clothes and Mathew Williamson was not familiar with menswear, for example.

7 Furthermore, they have introduced other limited collections, such as the environmentally friendly "H&M

Garden Collection". "H&M Fashion Against AIDS Festival Collection 2010" is the latest such collection that aims to bring awareness to the spread of AIDS to festivals. Or maybe it just wants to generate favorable publicity for H&M and link it to a humanitarian cause to avoid the sweatshop discussion…you choose. In any case the collection is pretty cool, and promoting safe sex is never bad. It has everything you would need for the rocky UK festival experience; I could definitely see Kate Moss strutting around in this in the mud. Except for clothes and accessories, it includes tents and sleeping bags. Lou Dillon and Lizzy Jagger excel as models in photos with a young rebellious attitude. The collection was just released on May 20 and 20 percent of sales will, according to H&M, go to fighting AIDS.

8 So tents now, what's next? Well, Zara has a home decorating brand as well. Of course Inditex (the Spanish holding company of Zara) could have chosen to start another completely separate furnishing brand, but they chose the Zara one. H&M has said their long-term strategy is diversifying into an IKEA or Wal-Mart concept. Hmmm ... maybe a Roberto Cavalli leopard grill or a Karl Lagerfeld black and white cart, why not? With H&M going into computer games with The Sims, this is not as far-fetched as it might sound ...

(1009 words)

New Words

credibility /ˌkredə'biləti/ *n.* 可靠性，确实性
collaboration /kəˌlæbə'reiʃ(ə)n/ *n.* 合作，协作；合著
comply /kəm'plai/ *v.* 应允，答应，依从，同意；遵照
exclusivity /ˌiksklu:'sivəti/ *n.* 排外，排外主义；排他性，排他主义
exterior /ik'stiəriə/ *n.* 外部，外面，表面，外形，外观
far-fetched /ˌfɔ:'fetʃt/ *adj.* 强词夺理的，牵强附会的
grill /gril/ *n.* （炉具的）烤架，焙盘
high-fashion /ˌhai'fæʃ(ə)n/ *adj.* （尤指女式服装）最新款式的，最新式样的
humanitarian /hju:ˌmæni'teəriən/ *adj.* 人道主义者的，慈善家的
hysteria /hi'stiəriə/ *n.* 癔症；（特指女人的）歇斯底里；病态的兴奋
inventory /'invəntəri/ *n.* 盘存，存货；（财产等的）清单，报表；（商品的）目录
legitimacy /li'dʒitiməsi/ *n.* 合法性；正统性；嫡系
mentality /men'tæləti/ *n.* 脑力，智力；精神；心理，意识；思想
oracle /'ɔrək(ə)l/ *n.* 神使，先知，预言者；大智者；（谑）圣人，哲人
outfit /'autfit/ *n.* 装置；（一定场合下穿的）全套衣装
pretension /pri'tenʃ(ə)n/ *n.* 虚饰，假装，做作；虚荣；自负
procurement /prə'kjuə(r)mənt/ *n.* 采购

send-off /'sendɔf/ n.　送别，送行
shoot /ʃu:t/ n.　摄影，拍电影
sketch /sketʃ/ n.　草图，粗样，略图；素描
sweatshop /'swetʃɔp/ n.　（非正式）血汗工厂
tag /tæg/ n.　标签；附笺，贴纸
underground /'ʌndəˌgraund/ adj.　反传统的，反现存体制的，（艺术等）先锋派的

Phrases and Expressions

holding company	控股公司
market specialist	市场专员
pair up with	（使）成对手；（使）成搭档
rip off	偷窃；扯掉；欺诈；剥削

Key Sentences

1. The people you would least expect had a fashion blog, took pictures of themselves in "Today's Outfit," and had opinions on coming trends, underground designer wear and new fashion shoots.

你最料想不到的人却拥有一个时尚博客，在"今日着装"上刊登他们自己的照片，同时对于未来的趋势、前沿设计师的穿着和崭新的时尚摄影有独到的见解。

2. H&M with their designer collaborations, and Zara with their incredible speed, participate in ripping off new designer styles.

在他们设计师的合作下的H&M和以令人难以置信的速度发展的Zara，双双拿下新的设计风格。

3. That market specialists, procurement personnel, and sales people are all sitting next to each other in Spanish La Coruña, allows the process between sketch through design to send-off to be instant; Zara can take a design from the drawing board to stores in only two weeks.

营销专家、采购人员和销售人员都在西班牙拉科鲁尼亚附近，这使得草图从设计到发出的过程变得即时；Zara可以在仅仅两周的时间里就能完成从图纸设计到商店销售的过程。

4. The Swedish retailer has paired up with high-fashion designers to create limited collections, creating hysteria among customers, media, and uncountable fashion-bloggers, selling out instantly.

瑞典的零售商配合高端时尚设计师制造数量有限的时装，让客户、媒体以及无数的时尚博主兴奋，迅速售罄。

5. "H&M Fashion Against AIDS Festival Collection 2010" is the latest such collection that aims to bring awareness to the spread of AIDS to festivals. Or maybe it just wants to generate favorable publicity for H&M and link it to a humanitarian cause to avoid the sweatshop

discussion.

最近的展览，如"H&M2010抗艾滋病日时装展"，旨在提高对艾滋病传播的意识，同时也能够制造对H&M有利的宣传，与人道主义密切关联，避免诸如"血汗工厂"之类的讨论。

Notes

1. Jimmy Choo is a British high fashion house specializing in luxury shoes, designer bags, and accessories. The company, J. Choo Limited, was founded in 1996 by couture shoe designer Jimmy Choo OBE and Vogue accessories editor Tamara Mellon OBE. In April 2001 Equinox Luxury Holdings Ltd. bought out Jimmy Choo's 50% ownership of the ready to wear business. The company was bought by current owners Labelux in 2011.

2. Elle is a worldwide magazine of French origin that focuses on women's fashion, beauty, health, and entertainment. Elle is also the world's largest fashion magazine, with 42 international editions in over 60 countries. It was founded by Pierre Lazareff and his wife Hélène Gordon in 1945. The title, in French, means "she".

3. La Coruña is a city and municipality of Galicia, Spain. It is the second-largest city in the autonomous community and seventeenth overall in the country. La Coruña is a busy port located on a promontory in the entrance of an estuary in a large gulf on the Atlantic Ocean. It provides a distribution point for agricultural goods from the region.

4. Karl Lagerfeld (born Karl Otto Lagerfeldt on September 10, 1933 in Hamburg) is a German fashion designer, artist and photographer based in Paris. He has collaborated on a variety of fashion and art related projects, most notably as head designer and creative director for the fashion house Chanel. Lagerfeld has his own label fashion house, as well as the Italian house Fendi. In 2004, he collaborated with H&M to launch the "Karl Lagerfeld for H&M" collection.

5. Roberto Cavalli (born 15 November 1940) is an Italian fashion designer from Florence. He is the first designer to figure out how to print on leather, changing rock star fashion overnight and making himself famous the world over. Known for his animal print designs and flowing long dresses, he is considered by many to be the new Gianni Versace. In 2007, he collaborated with H&M to create "Robert Cavalli at H&M" collection.

6. The Sims is a strategic life-simulation video game developed by Maxis and published by Electronic Arts. Its development was led by game designer Will Wright, also known for developing SimCity. It is a simulation of the daily activities of one or more virtual persons ("Sims") in a suburban household near SimCity.

7. This article is adapted from http://blog.modelmanagement.com/2010/06/01/zara-and-hm-fast-fashion-on-demand/#comments.

Post-Reading Exercises

- **Reading Comprehension**

Directions: *Read the passage and answer the following questions.*

1) What is the major difference between Jimmy Choo and H&M?

2) What are the different marketing strategies used by H&M and Zara respectively?

3) What kind of customer mentality did Zara create with the help of its marketing strategy? What about H&M?

4) How many stores has Zara owned internationally till now? What about H&M?

5) What's the writer's attitude towards H&M's future plan?

- **Translation**

Directions: *Translate the following sentences from Chinese to English, using the expressions in the brackets.*

1）学生统一穿校服有利于培养学生的团队精神。(That ... helps)

2）在比赛中，他被安排与一个得力的年轻人一起执行任务。(pair up with)

3）理论来源于实践，同时也为实践服务。(originate from)

4）这次活动旨在引起社会对全球变暖所带来的极地融化等危机的重视。(bring awareness to)

5）据报道该地区所有不服从安全监管的服装厂都将于近期关闭。(comply ... with)

Unit 8　International Trade of Textile and Fashion

PART ONE　Warm-up Activities

Chinese-made Textiles

New Words

tariff /'tærif/ *n.* 关税	restriction /ri'strikʃən/ *n.* 限制
correspondent /ˌkɔri'spɔndənt/ *adj.* 相应的	

Directions: *Listen to the passage and fill in each blank with the information you get from the recording.*

1. China declared its sharply increased export duties on Chinese-made textiles, But Washington and the European Union have more _____ on Chinese exports, which China said are unfair and _____ .

2. Minister of Commerce Chen Demin told a news briefing yesterday that _____ tariffs on 81 kinds of textile products will be _____ , including the 74 for which 400 per cent increases were announced. The latest restrictions _____ by the US side will affect US$2 billion worth of Chinese exports and 160,000 jobs, while the EU _____ will lead to a _____ of US $300 million exports and corresponding jobs.

3. So we have to make corresponding _____ adjustment since the EU and the US have set _____ on Chinese textile exports, said Chen.

4. We must be fair to Chinese _____ .

PART TWO Reading Activities

Passage 1 Eastern Players in the Global Fibres Industry

1 On a frosty February afternoon in the workroom of a northern Italian fabric R&D office outside Bergamo, the mood was quiet. Outside, winter fog nearly eclipsed Roncola Mountain in the Alps. People carried swatches of denim in blue, gray, ocher, pink. They brought color cards, market brochures, style specification sheets, or simply greetings. They were mostly the sales force of the company, usually spread out in more than twenty countries and five continents, brought partially together in anticipation of the season's fabric shows, the most important of which just a week away in Paris, with a smaller show in Milan in the days beforehand. The company, Legler, is one of Italian oldest textile firms. The head designer, Pascal Russo, has always worked twelve- or fourteen-hour days leading up to the shows, but this year he was surprised to nd he had time on his hands. He was relaxed, leaning back in a leather chair with his legs crossed in jeans and a powder-blue sweater. Pascal had been to these shows dozens of times and approached them with the same level of quiet confidence that an aging rock star takes the stage. For Pascal, there were more important things to worry over than the shows.

2 News that Armani, Gucci, and Prada were moving their manufacturing to Asia had shaken the industry in Italy, Pascal said. The country's high-end designer manufacturing has undergone such a profound change in recent years that it seems traumatized. "Gucci, Armani, Prada, these are Italian," he said. The word Italian emerged as an appeal, as if he was stating not what they were, but what he yearned for them to be. A day earlier Pascal's boss, Sergio, had called such moves "unacceptable for the Italian mentality."

3 Pascal appreciates ironic humor. When something strikes him as funny, he often does not laugh, but instead says, "This is funny." When something strikes him as wrong – a political situation, a rising crime rate, or a shortage of cheese or chocolate – he will say, "It is a pity." Often he describes unexpected places or things or events as "beautiful": the opening ceremony of the Turin Olympics, a juicer his wife bought, or the belongings found by an archeologist of an early hominid. They were all, in his aesthetic, "beautiful." Also beautiful is just about anything bearing a "Made in Italy" label, which means Pascal's world gets slightly less beautiful each day.

4 Italians had long taken pride in those things they designed and made. Even while Americans and Britons were demanding cheap Chinese-sourced goods a decade or more ago, Italians still had the majority of their consumer products made in their own country-particularly when it came to fashion. This is no longer the case for Italians, particularly in 2005 and 2006. Imports of denim jeans

rose by 260 percent between 2000 and 2005. The sheer number of "Made in China" labels hanging from the racks in stores from Milan to the Marché had the entire design and manufacturing industry in cultural and psychological chaos. It had always been believed that no one but an Italian could make Italian clothes, in the same way Italians are unwilling to accept the fact that anyone other than an Italian can make an Italian shoe. "To see 'Made in China' on them is...," Pascal struggled with the right words, "It is a pity. 'Made in Italy.' This still means something."

5 Pascal said Italy began to feel the consequences of the lifting of the quota system almost immediately. A Legler executive said the industry needed restructuring and less production. He talked about the English company Marks & Spencer, how everything they made used to come only from the UK. Now he said, "no one in the UK cares where M & S makes their textiles. I'm sure it'll go the same way in Italy."

6 Sergio said sometimes a company had to sell fabric without a profit to keep volume these days. "Volumes plummeted quicker than we expected. In 2005, compared to 2004, denim imports (to the European Union) increased something like two thousand percent," he said. He insisted though, that while manufacturing may disappear, styling would always remain in Italy. "Italy was one of the last countries to understand globalization and change. Finland and Sweden lost their manufacturing long ago, but they moved to technology, like Nokia. H&M also understood globalization. In Italy we are the tail of Europe because we still focus on manufacturing. On hardware."

7 Perhaps the most profound effect, though, was not the downturn in sales for companies like Legler and its competitors, but the inevitable layoffs. The main office had the uncomfortable task of cutting sixty jobs by the end of 2006. That's roughly 10 percent of the entire work force. Pascal was supposed to cut two from his department. It would mean a 30 percent drop in employment for his department, and it frankly didn't make sense to him. "If we could do the job with only four, why not have four all these years?" he asked. "From six to four? It's not possible." Italy is unique in that people tend to stay in their jobs; the newest employee on Pascal's team had been there nine years; the oldest, nearly thirty. Partly it is the Italian culture to stay at a job for one's entire life. If you had job security, then you stayed. But these days, it was also a lack of viable options.

8 But it also wasn't possible that things could stay as they were. In Italy, technical considerations were important in deciding who stayed and who went, but also taken into account were family situations and length of employment. Even if a five-year employee was the best in the department, she could still lose her job to a mediocre worker with a family who'd been there thirty years. It was the sort of civil law that seemed fair and maybe even progressive to someone staring down forced retirement, but in reality created multifarious real-life issues of fairness and egalitarianism.

(1006 words)

New Words

appeal /ə'pi:l/ *n.* 吸引力；*vt.* 恳求；呼吁
appreciate /ə'pri:ʃieit/ *vt.* 欣赏；领会
archeologist /ˌɑki'ɔlədʒist/ *n.* 考古学家
chaos /'keiɔs/ *n.* 混乱，无秩序
consequence /'kɔnsikwəns/ *n.* 后果，结果
denim /'denim/ *n.* 粗斜纹布，（复数）工作服，牛仔裤
downturn /'dauntə:n/ *n.* 低迷时期
eclipse /i'klips/ *v.* 使……黯然失色
executive /ig'zekjutiv/ *n.* 行政主管
frosty /'frɔsti/ *adj.* 严寒的，霜冻的
gray /grei/ *adj.* 灰色的，暗淡的
hominid /'hɔminid/ *n.* 原始人类
ironic /ai'rɔnik/ *adj.* 讽刺的，说反话的
juicer /'dʒu:sə/ *n.* 榨果汁器
layoff /'leiɔ:f/ *n.* 临时解雇，停止操作
lean /li:n/ *v.* 倾斜，倚靠
lift /lift/ *v.* 正式取消，结束
mentally /'mentəli/ *n.* 精神力，心理状态
ocher /'əukə/ *n.* 赭色
option /'ɔpʃən/ *n.* 选择
partially /'pɑ:ʃəli/ *adv.* 部分地
plummet /'plʌmit/ *vi.* 垂直落下，暴跌
profound /prə'faund/ *adj.* 深远的，深厚的，深重的
psychological /saikə'lɔdʒikəl/ *adj.* 心理的
quota /'kwəutə/ *n.* 配额，定额，限额
rack /ræk/ *n.* 行李架，衣架
restructure /ri:'strʌktrə/ *v.* 重组
sheer /ʃiə/ *adj.* 全然的
swatch /swɔtʃ/ *n.* （布料等的）样品
traumatize /'traumətaiz/ *vt.* 使受损伤
viable /'vaiəbl/ *adj.* 能生存的，可行的
yearn /jə:n/ *v.* 渴望，盼望

Phrases and Expressions

lead up to...　　　　　　　　　　　　　　就在……之前
on someone's hands　　　　　　　　　　由某人支配
powder blue　　　　　　　　　　　　　浅灰蓝色

Key Sentences

1. They were mostly the sales force of the company, usually spread out in more than twenty countries and five continents, brought partially together in anticipation of the season's fabric shows...
他们主要是公司的销售人员，通常分散在五大洲二十多个国家，有一些现在汇聚在一起等待这一季的织物展示……

2. The head designer, Pascal Russo, has always worked twelve- or fourteen-hour days leading up to the shows, but this year he was surprised to find he had time on his hands.
首席设计师帕斯卡·鲁索（Pascal Russo）在展示会前的日子里总是每天工作12到14个小时。但是今年他很惊讶地发现自己手头很有时间。

3. Pascal had been to these shows dozens of times and approached them with the same level of quiet confidence that an aging rock star takes the stage.
帕斯卡参加这种展示会已经有几十次了，他就像一个老练的摇滚明星上台一样沉静自信。

4. ... , as if he was stating not what they were, but what he yearned for them to be.
……，似乎他陈述的不是他们是什么，而是渴望他们会是什么。

5. When something strikes him as funny,...
当有什么东西让他觉得有趣，……

6. ...Italians still had the majority of their consumer products made in their own country—particularly when it came to fashion.
……意大利人大多数的消费品还是在自己国家生产——尤其是时装。

7. The sheer number of "Made in China" labels hanging from the racks in stores from Milan to the Marché had the entire design and manufacturing industry in cultural and psychological chaos.
从米兰到马尔什（Marché），商店衣架上到处可见"中国制造"的标牌，仅仅这个标牌的数量就已经搅乱了整个设计和制造行业的文化和心理。

8. Pascal said Italy began to feel the consequences of the lifting of the quota system almost immediately.
帕斯卡说意大利几乎马上就感受到取消配额制带来的后果。

9. Sergio said sometimes a company had to sell fabric without a profit to keep volume these days.
塞尔吉奥（Sergio）说这些日子，公司有时候为了销量，不得不无利润地销售织物。

10. If you had job security, then you stayed. But these days, it was also a lack of viable options.

如果工作使人有安全感，人们会留下来工作。但是这些日子，留在工作位置上也是由于缺乏其他可行的选择。

Notes

1. Bergamo：意大利城市，位于米兰东北40公里。
2. Alps：阿尔卑斯山。
3. Marks & Spencer（英国玛莎百货集团）是英国最大的商业集团，是跨国零售企业集团，创始于1897年，目前在全球都设有商店，具有很好的经营效益。在其成功的经验中很重要的一点是能从顾客的需要出发，主动开发产品。但玛莎集团并不自己投资建厂，而是将所涉及的产品交由制造商生产，被称为"没有工厂的制造商"。如在第一次世界大战期间，许多妇女进入工厂或作坊工作，产生了穿着轻便服装的需求，而一直没有服装厂大量生产和提供这样的服装。玛莎根据市场的需求，主动设计和开发了这类服装，并指导制造厂家大批量生产，向市场提供了品质优良、手工精巧、价格实惠的女式轻便服，满足了市场的需要，从而使公司获得了相应的利益，这在商业的品牌建立和拓展中是独树一帜的。

4. H&M 全称是 Hennes & Mauritz，1947年由埃林·佩尔森（Erling Persson）在瑞典创立，销售服装、配饰和化妆品。公司以销售量为衡量标准，兼顾流行、品质及价格，是欧洲最大的服饰零售商。总部在斯德哥尔摩，公司重要的职能部门如设计采购部、金融部、财务部、发展部、展示设计部、广告部、公关部、人事部、物流部、IT与客户服务部都设在总部。同时，公司在全球设有15个办事处，22个生产办公室，负责与亚洲和欧洲超过700家独立供应商合作。这22个生产办公室中，有9个在欧洲，11个在亚洲，1个在中美洲，1个在非洲。H&M时尚年度分为两季：春夏和秋冬。采购活动与市场导向一致，并根据分部在世界各地的销售点提供的数据，比如容易销售的商品、气候差异以及购物喜好等不断做出调整，使时尚流行的准确性

得到最大优化。商品筹备时间从 2、3 周到 6 个月不等，主要由商品的属性决定，正确的筹备时间是在价格、时间与品质方面保持平衡。

Post-Reading Exercises

- **Reading Comprehension**

Directions: *Read the passage and answer the following questions.*

1) What were the people busy with in that northern Italian fabric R&D office outside Bergamoa?

2) Was the head designer, Pascal Russo, working overtime this year for the shows?

3) What did Pascal think of the change happening in the field of high-end designer manufacturing in Italy?

4) According to the passage, why was Pascal's world getting slightly less beautiful each day?

5) What did "the Italian mentality" (in the last sentence of paragraph 2) mean?

6) Pascal said that it was a pity to see "Made in China?" According to the passage, what did he mean?

7) What were the consequences of lifting the quota system in Italy?

8) What did Sergio think of Italian textile manufacturing when he compared Italy with other EU countries?

9) How many employees were there in Pascal's department?

10) List some elements which may decide whether people will be laid off or not.

- **Vocabulary**

Directions: *Read the following groups of sentences carefully and discuss with your partner how the same word is used with different meanings in each group. Then translate them into Chinese.*

1) lift

    a. He stood there with his arms lifted above his head.

    b. The mist began to lift.
2) sheer
    a. A common use of sheer fabric is in curtains, which allows for sunlight to pass through during daylight while maintaining a level of privacy.
    b. I met her by sheer chance.
3) plummet
    a. Share prices plummeted to an all-time low.
    b. The great ship groped her way toward the shore with plummet and sounding line.
4) appeal
    a. The police made an appeal to the public to remain calm.
    b. The Beatles have never really lost their appeal.
    c. She appealed to the Supreme Court against her sentence.
5) approach
    a. The time is fast approaching when we shall have to make a decision.
    b. I'd like to ask his opinion but I find him difficult to approach.
    c. The school has decided to adopt a different approach to discipline.
6) move
    a. We were moved to tears by her story.
    b. I move that a vote be taken on this.
7) strike
    a. It suddenly struck us how we could improve the situation.
    b. The union struck for better work conditions.
8) thread
    a. The little girl threaded the shells together and wore them round her neck.
    b. I'm afraid I've lost the thread of your argument.
    c. This robe is embroidered with gold thread.
    d. A thread of light emerged from the keyhole.
9) appreciate
    a. You can't really appreciate foreign literature in translation.
    b. I would appreciate any comments you might have.
    c. Their investments have appreciated over the years.
10) industry
    a. Success comes with industry.
    b. Should industry be controlled by the state?
    c. His new play is a satire on the fashion industry.

- **Translation**

Directions: *Translate the following sentences into English, using the expressions in brackets.*

1）美国在过去30年里，纺织和服装行业工人的数量从240万直线下降到70万。（plummet）

_____

2）有些国家的纺织业经历了从传统到更先进的纺织品生产这个巨大的转变。（undergo）

_____

3）毕业前的几周对我来说最忙碌了。（lead up to）

_____

4）仅仅通过增加进口税来保护本国经济，这样做有意义吗？（make sense）

_____

5）在请她来跟我们合伙前我们得先考虑她过去的经历。（take into account）

_____

Passage 2 Weaving New Markets

1 The functionality of textiles has possibly been extended the furthest by the recent introduction of wearable electronics through the combination of nanotechnology, information technology and electronics in clothing. Apparel manufacturers have been able to use e-textiles to incorporate electronic links in their products to mobile phones, iPods and personal digital assistants.

2 Motorola has teamed up with "*Burton Snowboard*" of the US to launch a snowboarding jacket, helmet and beanie hat, which can be connected to a mobile phone and/or iPod via a control module on the jacket sleeve. Stereo speakers and microphone have been built into the hood and collar of the jacket.

3 The Burton e-textile product comprises a conductive fabric. Other e-textile innovations, based on conductive yarns, can illuminate apparels through light-emitting diodes. Thermochromic colours in fibres can also change the appearance of clothes in higher temperatures.

4 Within the speciality fibres sector, major advances have been achieved with high performance fibres, particularly those made from aramid, polyamide and carbon. These are now establishing themselves in a wider range of end-use markets.

5 An example of moves to bolster the potential of declining textile manufacturing areas of Europe is a research scheme in Northern Ireland, which has had a linen and wool weaving industry for more than two centuries. The project, sponsored by the UK government's Department of Trade and Industry and centred on the application of carbon fibre in the aerospace sector, involves the University of Ulster's leading research centre in 3D woven fibre preforming.

6 "We are taking textile weaving techniques and applying them to aerospace engineering technology to develop the potential to weave components for the next generation of aircraft and aircraft engines," says Justin Quinn, director of the university's engineering composites research centre.

7 The university believes the research will help to revive Northern Ireland's textile industry, which has been hit by a severe downturn, and also assist other parts of the UK's textile segment.

8 "We are one of the very small number of people in the UK who can weave these complex woven architectures on traditional weaving machinery," Quinn continues. "By transferring this technology into the aerospace industry we will create a productive relationship between the textile and aerospace industries."

9 The transformation of sections of the textile industry in the major industrialised countries is exemplified by the restructuring of leading fibre producers like DuPont and Teijin, both of which have global operations and are investing heavily in high-performance fibres.

10 In February, Teijin of Japan announced medium-term plans to raise the sales of its synthetic fibres business by 21% and to more than double its operating profit in the three years prior to 2008.

This will be achieved by concentrating capital and R&D investment on the growth activities of para-aramid, carbon and polyethylene naphthalate fibres while it is reorganising its commodity-type polyester fibres operations.

11 The key growth markets for its fibres will be automobiles, aircraft, information and electronics, healthcare and environment, and energy. In the automobile market it already provides fibres for use in 16 different parts or components of vehicles, ranging from filters and air-conditioner hoses to brake pads, steering wheel covers and seat sensors.

12 In July, Teijin is to phase in a 20% increase in the production capacity, to 23 000 t/year, of its Twaron para-aramid fibre at Emmen and Delfzijl in the Netherlands. It will be the third capacity increase of the fibre since it acquired the business from Akzo Nobel five and half years ago.

13 The fastest growing application areas for Twaron, which is produced as a yarn and as pulp, are ballistics and friction, such as brake pads and gaskets. It is also increasing sales in rubber reinforcements, optical fibre cables, composite materials and civil engineering/geotextiles.

14 After selling its $6 billion business in bulk fibres and intermediates to Koch Industries two years ago, DuPont has been channeling a lot of resources into the expansion of its high performance and advanced fibres. DuPont's safety and protection segment, a high proportion of whose sales are in these fibres, is one of the most profitable in the company.

15 A version of its non-woven Tyvek textile made of very fine, high-density polyethylene fibres has recently been launched in Europe as an energy efficient vapour-permeable roofing material. Kevlar, a low-weight aramid fibre five times stronger than steel, discovered 40 years ago, has just been re-engineered to provide residents protection from the dangers of hurricanes and tornadoes.

16 DuPont last year increased its equity stake in the start-up company, Magellan Systems International, based at Richmond, Virginia, only a few miles from the site of DuPont Advanced Fiber Systems. Magellan has developed a polymer fibre which is even stronger and lighter than Kevlar.

17 A new direction for DuFont's fibres operation has been developing advanced fibres made from crop-based materials. At a plant in Kinston, North Carolina, it is making from corn-derived 1.3-propanediol (PDO) Sorona, a fibre for apparel and industrial applications.

18 The development by DuPont in the 1990s of a process for a commercially viable bio-PDO to make a smart fibre with shape-recovery properties such as Sorona perhaps would not have happened without the emergence of highly competitive textile manufacturers in the low-cost production areas of Asia.

(854 words)

New Words

aramid /ˈærəmid/ *n.* 芳香族聚酰胺

ballistics /bə'listiks/ *n.* 导弹学，发射学
beanie /'bi:ni/ *n.* 小便帽
bolster /'bəulstə/ *n.* 长枕，靠枕
bulk /bʌlk/ *adj.* 大量的，散装的
capital /'kæpitl/ *n.* 资本；首都
component /kəm'pəunənt/ *n.* 元件，组成部分
composite /'kɔmpəzit/ *n.* 合成物，复合材料
conductive /kən'dʌktiv/ *adj.* 传导性的
digital /'didʒitəl/ *adj.* 数字的，数码的
diode /'daiəud/ *n.* 二极管
exemplify /ig'zemplifai/ *vt.* 例证，示范
filter /'filtə/ *n.* 滤色镜，过滤器
functionality /ˌfʌŋkʃə'næliti/ *n.* 功能，功能性
gasket /'gæskit/ *n.* 垫片
hurricane /'hʌrikən;'hə:ˌkein/ *n.* 飓风
high-performance *adj.* 高性能的
hood /hud/ *n.* 头巾，兜帽
hose /həuz/ *n.* 水管，橡皮软管
illuminate /i'lju:mineit/ *vt.* 照明；阐释，说明
incorporate /in'kɔ:pəreit/ *vt.* 合并
Kevlar *n.* 纤维 B（一种质地牢固重量轻的合成纤维）；［商标］凯夫拉尔
launch /lɔ:ntʃ/ *vt.* 发起
linen /'linin/ *n.* 亚麻布，亚麻制品
module /'mɔdju:l/ *n.* 组件，单元，模块
nanotechnology /ˌnænətek'nɔlədʒi/ *n.* 纳米技术
optical /'ɔptikəl/ *adj.* 光学的
para-aramid *n.* 对位芳纶
phase /feiz/ *n.* 阶段，时期；*v.* 逐步执行
polymer /'pɔlimə/ *n.* 聚合体
polyamide /pɔli'æmaid/ *n.* 聚酰胺，尼龙；*vt.* 支持，鼓励
polyethylene /ˌpɔli'eθəli:n/ *n.* ［离分子］聚乙烯
preforming /pri:'fɔ:miŋ/ *n.* 预成型，压片
pulp /pʌlp/ *n.* 果肉；纸浆
revive /ri'vaiv/ *vt.* 唤醒，使……重生
scheme /ski:m/ *n.* 计划，方案
segment /'segmənt/ *n.* 部分

sensor /'sensə/ *n.* 传感器，探测器
thermochromic /'θə:məu'krɔmik/ *adj.* 热色性的
tornado /tɔ:'neidəu/ *n.* 龙卷风
transfer /træns'fə:/ *vt.* 转移，调任
vapour /'veipə/ *n.* 水蒸气，雾气

Phrases and Expressions

equity stake   股本，股份
muscle in   硬挤进

Key Sentences

1. The functionality of textiles has possibly been extended the furthest by the recent introduction of wearable electronics through the combination of nanotechnology, information technology and electronics in clothing.
   最近推出可穿戴的电子服装，结合了纳米技术、信息技术和电子技术，纺织品的功能性有可能已经延伸到最远了。

2. Stereo speakers and microphone have been built into the hood and collar of the jacket.
   立体声音响系统和麦克风已经植入外套的兜帽和领子。

3. ..., can illuminate apparel through light-emitting diodes.
   ……，通过发光二极管，能照亮衣服。

4. Within the speciality fibres sector, major advances have been achieved with high performance fibres...
   在专业纺织品领域，主要的发展是高性能纺织品的出现……

5. The university believes the research will help to revive Northern Ireland's textile industry, which has been hit by a severe downturn, and also assist other parts of the UK's textile segment.
   阿尔斯特大学相信，这项研究将有助于恢复在严重低迷时期受到打击的北爱尔兰纺织品行业，而且对英国纺织品部门的其他方面也有帮助。

6. The transformation of sections of the textile industry in the major industrialised countries is exemplified by the restructuring of leading fibre producers like...
   在主要工业国进行的纺织品行业部门转型是以像……这类的主要织物生产商重组为范例的。

7. The key growth markets for its fibres will be automobiles, aircraft, information and electronics, healthcare and environment, and energy.
   织物的主要增长市场将会是汽车、飞机、信息和电子、健康和环境以及能源。

8. DuPont has been channeling a lot of resources into the expansion of its high performance and advanced fibres.

杜邦公司已经把很多资源用于扩展高性能和先进的织物。

9. A version of its non-woven Tyvek textile made of very fine, high-density polyethylene fibres has recently been launched in Europe as an energy efficient vapour-permeable roofing material.

一种由高密度聚乙烯纤维制成的非常优质的非织造布 Tyvek 纺织产品最近投放在欧洲市场，作为节能透气的屋顶材料。

10. ...perhaps would not have happened without the emergence of highly competitive textile manufacturers in the low-cost production areas of Asia.

如果没有出现来自亚洲地区纺织品生产商低成本产品的高度竞争，……也许不会发生。

Notes

1. E-textile also known as electronic textiles or smart textiles, are fabrics that enable computing, digital components, and electronics to be embedded in them.

2. Thermochromic (of a substance) undergoing a reversible change of color when heated or cooled.

3. Research and Development (R&D) investigative activities that a business chooses to conduct with the intention of making a discovery that can either lead to the development of new products or procedures, or to improvement of existing products or procedures. Research and development is one of the means by which business can experience future growth by developing new products or processes to improve and expand their operations. While R&D is often thought of as synonymous with high-tech firms that are on the cutting edge of new technology, many established consumer goods companies spend large sums of money on improving old products. For example, Gillette spends quite a bit on R&D each year in ongoing attempts to design a more effective shaver.

4. Twran is a very strong, light para-aramid fiber (poly-paraphenylene terephthalamide) developed and produced exclusively by Teijin Aramid. It has a high modulus, and is thermally stable, and highly impact and chemical resistant. Since its development in the 1960s and 1970s, Teijin Aramid is producing the monomer and polymer in Delfzijl and the yarn in Emmen (The Netherlands). Twaron is used in a huge range of specialist applications, specifically customized to complement our customers' products, and is widely recognized by many industries as a quality product with great durability and recycling potential.

5. Akzo Nobel（阿克苏诺贝尔）是财富 500 强公司之一，总部在荷兰，主要经营化学品。

6. Tyvek 是杜邦公司优秀的非织造产品，由高密度聚乙烯纤维制成，具有均衡的物理特性，有厚度薄、重量轻、不易变形、柔软平滑、坚韧、抗撕裂、不透明、抗水渍、表面摩擦力小、弹性大等特点，结合了纸、布和薄膜的优点。Tyvek 是一种用途极为广泛的材料。

Post-Reading Exercises

• **Reading Comprehension**

Directions: *Read the passage and decide whether the following statements are true(T) or false(F) and give reasons.*

1) _____ The snowboarding jacket launched by Motorola comprises a conductive fabric.

2) _____ High performance fibres are the fibres which are made from aramid, polyamide and carbon.

3) _____ The University of Ulster is doing a project which is supposed to be able to revive Northern Ireland's textile industry.

4) _____ The research done in the University of Ulster is to weave carbon fibre on innovative machinery.

5) _____ DuPont has developed more than one fibre which is stronger and lighter than steel.

• **Vocabulary**

Directions: *Complete the following sentences with the proper forms of the words given in the box.*

| launch | transfer | illuminate | capital |
| digital | exemplify | segment | revive |

1) This article mostly _____ the importance of brand innovation.

2) To _____ what I mean, let us look at our annual output.

3) _____ resources can open up physical space in the library without diminishing the total content available.

4) The company researched the market demand and decided to _____ their new product.

5) Economists argued that freer markets would quickly _____ the region's economy.

6) With a conventional repayment mortgage (按揭), the repayments(还款) consist of both _____ and interest.

7) The exhibition was _____ to another city two days later.

8) The company dominates this _____ of the market.

• **Translation**

Directions: *Translate the following Chinese terms into English.*

1）碳纤维 _____

2）传导性纤维 _____

3）高性能纤维 _____

4）合成纤维 _____

5）智能纤维 _____
6）光导纤维 _____
7）电子织物 _____
8）非织造布 _____
9）节能的 _____
10）透气的 _____

Passage 3 Markets and Value in Clothing and Modeling

1 Defining high fashion clothing and describing the UK fashion retailing arena is far from easy. The fashion retailing industry can be dissected across a variety of dimensions, including size, ownership, and market segment. It is difficult to study this industry because it is fast changing, with newcomers often gaining market share very rapidly, and because it is characterized by diverse arrangements for production, marketing and sales. Furthermore, its relationship between the two sectors-multiple and independent designer-is neither simple nor stable.

2 Fashion clothing markets in the UK are highly differentiated, broken down into many different levels; in effect, different markets. These levels are defined partly by price-points, which is the term buyers use for the price level at which particular consumers are prepared to pay for particular items of clothing. The entry of low-priced retailers in the UK, such as New Look, Primark and Matalan, as well as the growth of super market fashion brands, such as "Florence and Fred" for Tesco and "George" for Asda, have taken the fashion retailing world by storm and the strength of these retailers continues unabated (Mintel 2002a,b). Besides the obvious appeal of low price, it is the success of these retailers in fast meeting trends that ensures their continued success. Indeed, the term "fast fashion" is often used to describe how stores, like Primark and other more expensive but equally popular high street stores like Zara and Topshop, can meet trends in season.

3 The significance of fast fashion has been accompanied by the continued charisma of individual designers and designer clothing at the other end of the market. Indeed, one characteristic in recent years, described by Crew and Forester(1993), is the polarization of fashion retailing in the UK over the 1990s between discounting outlets and an emerging designer sector, a polarity hinged upon cheap, mass-produced clothes on the one hand and unique, high-quality fashion clothing on the other. The contemporary UK High Street remains highly polarized between cut-price outlets, such as "Primark", and high fashion stores, blending "own ranges" and "designer", i.e. branded clothing, such as "Whistles" or "Joseph". As part of this polarization, the UK clothing sector has become concentrated in the hands of a small number of large multiples. Unlike other European countries, multiples make up larger proportion of the market in the UK with independent stores constitute a smaller percentage (Mintel 2002a,b)

4 The buyers I followed in Selfridges are firmly located at the "high" end of women's clothing, what is also referred to as "designer" clothing. The size of the UK designer fashion economy alone was estimated at £600 million in 1996(DCMS 1998, 2001), and more recently , as £800 million in 2006 by Mintel (in Roodhouse 2003). However, despite its apparent obviousness, the term designer clothing is surprisingly vague. Mintel (2002a) defines it very broadly to refer to four things: "couture", dominated by French-based international brands like "Dior" or "Chanel"; "international designers", referring to a label usually dominated by one name: "Donna Karan" or "CK"; "diffusion designers"

who produce "high-street" ranges for stores, such as Jasper Conran at Debenhams; and "high fashion", referring to new designers often endorsed by celebrities. When I pressed Julia the head buyer for Designer Wear to define this term she noted, "It is not a terribly specific term any more". However, she went on to say,

5 "the term does really indicate a certain level in the market, sort of the higher level of the market, which includes some of the big lifestyle brands like 'DKNY' where they sell everything from leisure wear, smart formal wear, evening wear, sleep wear, you know, they do the whole thing, that you could live your life in DKNY if you wanted to. So it encompasses that, but at the same time it does encompass the smallest designers as well. So don't get confused by the title, it doesn't mean anything!"

6 Thus, designer fashion for her, and all the buyers at Selfridges, sits within the "higher level" range identified by Mintel in 2002a, encompassing only the first two definitions, although we must set aside couture, as its market is economically insignificant (though symbolically important). Designer clothing at Selfridges refers to a number of things. Firstly, prêt-a-porter "collections" of individual designers that retail at considerable prices, are made to order and are therefore quite exclusive. The most elite designers do both prêt-a-porter and couture, such as "Givenchy" or "Yves Saint Laurent". Prêt-a-porter has its own spatial and temporal arrangements, showing at the international "collections", or "fashion weeks" in London, New York, Milan and Paris. This involves the physical movement of the worlds fashion players—journalists, buyers, designers, models, etc. - around the globe at the same time, with some shows also taking place in cities peripheral to this historic configuration of fashion cities but who now organize their own fashion weeks (Breward and Gibert 2006). These cosmopolitan hubs are not only where the design houses are located; they are the locations for all the machinery of styling, photographing and marketing. As I describe below, the collections set a temporal rhythm to designer fashion. These shows are bi-annual and this structures the flow of such clothing, with buying cycles geared to two main seasonal deliveries.

7 The temporal dimensions of female fashion at this level are also very particular. Prêt-a-porter is produced to order in smaller batches, unlike high street female clothing made in bulk, with a longer time lag between order and delivery. It is organized into two seasons. Autumn/Winter in January/February, and Spring/Summer in September/October. This means that the clothes are shown and purchased four to six months ahead of the actual (real-time) season, making for a rather slower fashion cycle than the contemporary high street which meets trends closer to real season. To address this, big brands and major designers now show "mid-season" or "pre-season" collections, such as the "Cruise" collection in October/November. Indeed, those brands that do pre-season collections account for 65 to 70 per cent of sales at Selfridges, hence greater even than mainline collections. However, most small to medium-sized designers do not have the capability (investment, manufacturing and time) to produce per-season collections so the bi-annual buying cycle remains

significant and sets the overall pace of fashion for the majority of designers.

(1094 words)

New Words

appeal /ə'pi:l/ *n.* 呼吁，吸引力；*v.* 恳求，有吸引力
arena /ə'ri:nə/ *n.* 竞技场
batch /bætʃ/ *n.* 批次；*v.* 分批处理
characterize /'kærəktəraiz/ *v.* 赋予……特色
charisma /kə'rizmə/ *n.* 超凡魅力，领袖气质
configuration /kənˌfigə'reiʃn/ *n.* 结构，布局，形态
cosmopolitan /ˌkɔzmə'pɔlitən/ *n.* 世界主义者；*adj.* 世界性的
couture /ku'tjuə(r)/ *n.* 高级时装
differentiate /ˌdifə'renʃieit/ *v.* 区分
diffusion /di'fju:ʒn/ *n.* 扩散，传播
dimension /dai'menʃn/ *n.* 方面，维度
dissect /di'sekt/ *v.* 解剖，详细分析
diverse /dai'və:s/ *adj.* 多种多样的
encompass /in'kʌmpəs/ *v.* 围绕，包括，完成
endorse /in'dɔ:s/ *v.* 赞同，支持
exclusive /ik'sklu:siv/ *adj.* 排外的，唯一的，奢华的
gear /giə(r)/ *n.* 齿轮；*v.* 调整
hinge /hindʒ/ *n.* 铰链；*v.* 依情况而定
hub /hʌb/ *n.* 中心
multiple /'mʌltipl/ *adj.* 许多的；*n.* 倍数
peripheral /pə'rifərəl/ *adj.* 外围的，不重要的
polarity /pə'lærəti/ *n.* 两极
polarization /ˌpəulərai'zeiʃn/ *n.* 两极化
polarized /'pəuləraizd/ *adj.* 两极化的
prêt-a-porter /p'ret'eip'ɔ:tər/ *n.* 现成服装
proportion /prə'pɔ:ʃn/ *n.* 比例
retail /'ri:teil/ *n. vt.* 零售
spatial /'speiʃl/ *adj.* 空间的
temporal /'tempərəl/ *adj.* 时间的
unabated /ʌnə'beitid/ *adj.* 不衰退的，不减弱的
vague /veig/ *adj.* 模糊的；茫然的

Phrases and Expressions

account for	说明（原因、理由等）
hinge upon	依……而定
in effect	实际上；有效
set aside	搁置，不考虑
take...by storm	在……大获成功

Key Sentences

1. Indeed, one characteristic in recent years, described by Crew and Forester(1993), is the polarization of fashion retailing in the UK over the 1990s between discounting outlets and an emerging designer sector, a polarity hinged upon cheap, mass-produced clothes on the one hand and unique, high-quality fashion clothing on the other.

如同 Crew 和 Forester(1993) 所描述的一样，事实上，近几年的一个特点是英国在 20 世纪 90 年代时装零售业处于折扣店和新兴设计师部分的两极化。一端是廉价的批量生产的服装，另一端是独特高质量的时装。

2. Mintel (2002a) defines it very broadly to refer to four things: "couture", dominated by French-based international brands like Dior or Chanel; "international designers", referring to a label usually dominated by one name: Donna Karan or CK; "diffusion designers" who produce "high-street" ranges for stores, such as Jasper Conran at Debenhams; and "high fashion", referring to new designers often endorsed by celebrities.

Mintel(2002a) 广义地把它界定为四类："高级时装"，主要指法国本土的国际品牌，如迪奥或香奈儿；"国际设计师"，通常指以某人名字命名的品牌，如 Donna Karan 或者 CK；"大众设计师"为"高街"上的各店生产，比如在 Debenhams（英国知名百货公司）的 Jasper Conran；"御用时装设计师"，指的是受名人钟爱的新晋设计师。

3. These cosmopolitan hubs are not only where the design houses are located; they are the locations for all the machinery of styling, photographing and marketing.

这些世界性的中心地区不仅仅是设计室的所在地，它们还是造型、摄影和营销等所有机构的所在地。

4. Prêt-a-porter is produced to order in smaller batches, unlike high street female clothing made in bulk, with a longer time lag between order and delivery.

成衣是小批量订制，不像大批量生产的高街女时装，其订制和发货之间有相对较长的时间差。

Notes

1. New Look 创建于 1969 年，是英国时装零售巨头之一。信奉便宜即时尚（Cheap is Chic）。总部位于英国多塞特郡的维茅斯镇，目前在 23 个国家开设店铺，全球服务包括电子商务在内超过 120 个市场领域。Primark 是总部设在都柏林的爱尔兰服装零售商。该公司是英国食品加工公司 ABF 的附属公司。Primark 是英国大众化品牌，被冠以"街头进步最多的商店信""最实惠商店""50 英镑花费中最佳商店""市中心年度最佳商店""英国年度经销商"等称号。Primark 以百姓能承受的价格走"大众时尚路线"，成功带动欧洲时尚界"低消费高时尚"的新风，是老牌的英国服饰品牌，旗下拥有诸多适合不同年龄风格的品牌线。Matalan 是英国服装零售巨头，由成立者 John Hargreaves 的两个孩子 Matt 和 Alan 的名字命名。其商铺一般设址在郊区。Matalan 以难以置信的超低价格出售当季所有最新时装，包括顶级设计师的作品，同时售卖各种家用纺织品。

2. Florence and Fred & George Florence and Fred 是英国最大零售公司 Tesco 下属成衣品牌。George 则是沃尔玛英国子公司 Asda 旗下的一个时装品牌。

3. Whistles & Joseph 是英国伦敦 OASIS 旗下顶级品牌。凯特王妃在 2012 伦敦奥运会闭幕式上身穿一袭 Whistles 裹身裙亮相，让 Whistles 一夜成名。其服饰以皮夹克、羊绒制品和丝绸裙子为主，注重手工刺绣和钉珠等细节，精致而讲究。Joseph 也是英国本土服装品牌，曾获英国公民最喜爱的设计师品牌。

4. Selfridges 创立于 1909 年，是伦敦最著名的百货公司之一。该公司汇聚了数量众多的大众流行品牌及设计师专柜。

Post-reading Exercises

- **Reading Comprehension**

Directions: *Read the passage and decide whether the following statements are true(T) or false(F).*

1) _____ The fashion retailing industry in the UK is changing quickly.

2) _____ The low-priced fashion brands are not very popular in the UK market.

3) _____ There are more cut-price outlets than high fashion stores in the UK.

4) _____ New Look cannot be defined as designer clothing.

5) _____ Female clothing at Selfridges has faster fashion cycle than prêt-a-porter.

- **Vocabulary**

Directions: *Complete the following sentences with the proper forms of the words given in the box.*

| multiple | temporal | peripheral | appeal |
| diverse | endorse | batch | hinge |

1) His acceptance will _____ upon terms.
2) Friendship _____ joy and divides grieve.
3) India has always been one of the most religiously _____ countries.
4) We are still waiting for the first _____ to arrive.
5) This issue explores some of the _____ dimensions of art.
6) Companies are increasingly keen to contract out _____ activities like training.
7) I can _____ their opinion wholeheartedly.
8) The United Nations has _____ for help from the international community.

- **Translation**

Directions: *Translate the following Chinese terms into English.*

1) 时装经济 _____
2) 漂亮的正式着装 _____
3) 以不菲的价格零售 _____
4) 外围城市 _____
5) 设定时间节奏 _____
6) 小批量订制 _____
7) 订制和发布之间的时间差 _____
8) 为大多数设计师设定总体的时装更新速度 _____

Glossary

abandon *vt.*　抛弃	U6-3
Abydos *n.*　阿比多斯（埃及古城）	U3-2
accessory *n.*　饰品	U7-1
accredit *vt.*　信任，认可	U6-1
accusation *n.*　指责，指控，控告	U6-3
acid *n.*　［化］酸，酸性物质；*adj.*　酸的，酸性的，酸味的	U1-1；U2-1；U4-2
acquisition *n.*　收购，获得	U7-1
acrylic *n.*　丙烯酸纤维，腈纶	U1-1
adjustment *n.*　调节，调整	U6-1
adorn *vt.*　装饰，使生色（+with）	U1-1
advanced *adj.*　前进的，预先的，先进的，高级的	U7-1
aesthetic *adj.*　美学的	U3-1
affinity *n.*　亲和性	U2-3
affluence *n.*　富裕，富足，丰富	U1-3
aftermath *n.*　（战争、事故的）后果，创伤	U6-3
agrotextiles *n.*　农用织物	U1-1
airbrush *v.*　用喷枪喷；*n.*　喷枪	U6-3
align *vt.*　使结盟；使成一行；匹配；*vi.*　排列；排成一行	U2-3；U3-1；U7-1
alkali *n.*　碱；*adj.*　碱性的	U2-1
allergenic *adj.*　引起过敏症的，导致过敏的	U1-1
allusion *n.*　隐喻，典故；暗示，暗指	U1-2
amateur *n.*　业余从事者；外行人；爱好者；*adj.*　业余的；外行的	U6-1
anatomy *n.*　解剖，解剖学	U6-1
androgynous *adj.*　性别特征不明显的，中性的	U6-3
angora *n.*　安哥拉兔毛（或山羊毛）线（或织物）	U7-2
antiquity *n.*　古代，古物，古迹	U1-2
apologist *n.*　辩解者，辩护者	U6-3
apparel *n.*　服饰，服装	U2-2；U6-1；U7-1
appeal *n.*　呼吁，吸引力；*v.*　恳求，有吸引力	U8-1；U8-3
appreciate *vt.*　欣赏，领会	U8-1

aramid *n.*　芳香族聚酰胺　　　　　　　　　　　　　　　　　　　U8-2

archeologist *n.*　考古学家　　　　　　　　　　　　　　　　　　U8-1

arena *n.*　竞技场　　　　　　　　　　　　　　　　　　　　　　U8-3

arsenal *n.*　军械库，武器库　　　　　　　　　　　　　　　　　　U7-1

article *n.*　（物品的）一件，物品，商品　　　　　　　　　　　　U6-1

asbestos *n.*　石棉；*adj.*　石棉的　　　　　　　　　　　　U1-1；U2-1

assortment *n.*　各种各样　　　　　　　　　　　　　　　　　　　U1-1

attendee *n.*　参加者，出席者　　　　　　　　　　　　　　　　　U7-1

attire *n.*　服装，衣着，盛装　　　　　　　　　　　　　　　　　U6-1

authenticity *n.*　确实，真实性　　　　　　　　　　　　　　　　U1-3

aversion *n.*　讨厌，讨厌的人　　　　　　　　　　　　　　　　　U5-2

backing *n.*　支持，后援　　　　　　　　　　　　　　　　　　　U6-3

backlash *n.*　反击，后冲；激烈反应，强烈反响　　　　　　　　　U1-3

ballistics *n.*　导弹学，发射学　　　　　　　　　　　　　　　　U8-2

bang-on *adj.*　极精确的；非常有效的　　　　　　　　　　　　　U7-2

bank *n.*　系列，组，库　　　　　　　　　　　　　　　　　　　　U6-3

barre *n.*　纬向条花　　　　　　　　　　　　　　　　　　　　　U4-2

batch *n.*　批次；*v.*　分批处理　　　　　　　　　　　　　　　U8-3

be superseded by　为……取代　　　　　　　　　　　　　　　　　U3-2

beanie *n.*　小便帽　　　　　　　　　　　　　　　　　　　　　　U8-2

beforehand *adv.*　事先，预先　　　　　　　　　　　　　　　　U6-2

bib *n.*　（小儿）围涎，围嘴；围腰的上部　　　　　　　　　　　　U7-2

biography *n.*　传记，档案，个人简介　　　　　　　　　　　　　U5-2

blazer *n.*　（法兰绒的）运动上衣　　　　　　　　　　　　　　　U5-3

bleach *vt.&vi.*　使（颜色）变淡，变白；漂白（使）晒白，退色　　U1-1

bleach *n.*　漂白剂，漂白坯布，漂白工厂，漂白工人　　　　　　　U2-1

blemish *vt.*　有损……的完美，玷污　　　　　　　　　　　　　　U6-3

blouse *n.*　女衬衫　　　　　　　　　　　　　　　　　　　　　　U7-2

bobbin *n.*　纱筒　　　　　　　　　　　　　　　　　　　　　　　U3-3

bolster *n.*　长枕，靠枕　　　　　　　　　　　　　　　　　　　U8-2

boom *n.*　兴旺，繁荣　　　　　　　　　　　　　　　　　　　　U1-2

Bougras *n.*　布格拉斯　　　　　　　　　　　　　　　　　　　　U3-2

boutique *n.*　时装用品小商店，流行女装商店，时装精品店　　　　U6-2

brand name *n.*　商标，品牌名称　　　　　　　　　　　　　　　U6-2

brightness *n.*　明亮，鲜艳，鲜艳度，（色彩）明度　　　　　　　U2-1

budding *adj.*　萌芽的，开始发育（发展）的，初露头角的；*n.*　发芽，萌芽；

	vt.&vi. 动词 bud 的现在分词	U1-3
bulge *vi.*	膨胀，凸出，鼓起	U6-3
bulk *adj.*	大量的，散装的	U8-2
bulkiness *n.*	庞大，笨重	U5-1
bulky *adj.*	（bulkier 比较级）粗大的，大的	U3-1
burrs *n.*	毛边，过火砖（burr 的复数形式）；	
	v. 从……除去毛刺，在……上形成毛边	U3-3
bust *n.*	失败，破产，经济萧条	U1-3
Byzantine *n.*	拜占庭人，拜占庭式建筑；	
	adj. 拜占庭的，拜占庭式的；错综复杂的，诡计多端的	U1-2
cabaret *n.*	卡巴莱，有歌舞表演的夜总会；（餐馆、夜总会等处的）歌舞表演	U6-3
cake *vt.&vi.*	（使）结块；（使）胶凝；（厚厚一层干后即变硬的软东西）覆盖	U6-3
capacity *n.*	能力，容量，资格，地位，生产力	U5-1
capital *n.*	资本；首都	U8-2
cardigan *n.*	（开襟）羊毛衫，羊毛背心，开襟绒线衫	U7-2
carding *n.*	梳理，梳棉	U3-1
carrier *n.*	导纱器，横机机头	U5-1
cashmere *n.*	开司米，山羊绒	U2-2
celebrity *n.*	名人，名流	U6-3
chaos *n.*	混乱，紊乱	U6-3；U8-1
characteristics *n.*	特征	U3-1
characterize *v.*	赋予……特色	U8-3
charisma *n.*	超凡魅力，领袖气质	U1-3；U8-3
chimique *n.*	化学；*adj.* 化学的	U4-3
chino *n.*	斜纹棉布，斜纹棉布裤；*adj.* 用斜纹棉布做成的	U5-3
circa *prep.*	大约于；*adv.* 大约	U3-2
claim *vi.*	提出要求；*vt.* 要求，声称，认领；*n.* 要求，声称，索赔	U5-2
clarify *v.*	说明，讲清楚；阐明，澄清	U6-3
clergyman *n.*	牧师，教士	U5-2
cleric *n.*	牧师，教士；*adj.* 牧师的，教士的	U5-2
climatology *n.*	气候学、风土学	U2-1
clothe *vt.*	给……穿衣，为……提供衣服；覆盖，使披上 (+in)	U1-1
coarse *adj.*	粗俗的，粗糙的	U5-2
cocoon *vt.*	包围，包裹；*vi.* 作茧；*n.* 茧，茧状物	U1-3
coin *n.*	硬币，钱币；*vt.* 杜撰，创造，铸造（钱币）	U5-1
collaboration *n.*	合作，协作；合著	U7-3

collaborator *n.*	通敌者，合作者	U6-3
collage *n.*	拼贴画，拼贴艺术；杂烩；收集品；收藏品	U1-1
collarless *adj.*	无领的	U6-3
collection *n.*	作品集	U6-2
combing *v.*	梳理，精梳	U3-1
comply *v.*	应允，答应，依从，同意；遵照	U7-3
component *n.*	元件，组成部分	U8-2
composite *adj.*	混合成的，综合成的，复合的	U1-1
composite *n.*	合成物，复合材料	U8-2
conceive *vi.*	怀孕，设想，考虑；*vt.* 怀孕，构思，以为，持有	U5-2
concentration *n.*	浓度，浓缩，专心，集合	U5-1
conductive *adj.*	传导性的	U8-2
configuration *n.*	结构，布局，形态	U8-3
configure *v.*	使成形	U4-1
confine *vt.*	限制，紧闭，使局限；*n.* 范围，限制，约束	U1-2
congested *adj.*	拥挤的，堵塞的	U6-1
connotation *n.*	含义，言外之意，内涵	U1-3
consequence *n.*	后果，结果	U8-1
conservative *adj.*	保守的；（式样等）不时新的	U5-3
conspicuous *adj.*	显著的，明显的，突出的	U1-3
construction *n.*	组织结构	U5-1
contemporary *adj.*	当代的同时代的；*n.* 同时代的人，同时期的人	U5-2
convent *n.*	女修道院	U6-3
conventional *adj.*	传统的，符合习俗的，常见的，惯例的	U5-1；U5-3
cord *n.*	绳索；*vt.* 束缚，用绳子捆绑	U3-1；U4-3
corduroy *n.*	灯芯绒	U4-3
corset *n.*	（妇女）紧身胸衣	U6-3；U7-2
cosmopolitan *n.*	世界主义者；*adj.* 世界性的	U8-3
costume *n.*	戏服，民族服装，装束	U6-1
coupon *n.*	赠券，（连在广告上的）预约券，优待券，优惠券	U7-1
course *n.*	线圈横列	U5-1
couture *n.*	高级时装	U8-3
couturier *n.*	女装设计	U6-3
craft *n.*	手艺，工艺；技巧，技能	U6-2
Craft Yarn Council of America	美国手工纱线制品委员会	U3-1
crease *n.*	折痕	U2-3

crease *vt.* 使有皱褶，弄皱；表面被子弹擦伤或击伤；*vi.* 起皱	U5-3
credibility *n.* 可靠性，可信性	U6-2；U7-3
crochet hook 钩针	U3-1
crocheting *n.* 钩编，钩编工艺	U1-1
crunch *n.*（经济等）紧缩状态	U7-2
cuff *n.* 袖口，护腕	U6-3
cufflink *n.*（衬衣的）袖扣	U5-3
cult *n.* 狂热崇拜，迷信（对象）；异教，邪教	U1-3
cultivation *n.* 种植，栽培；教化，培养	U1-2
curate *n.* 助理牧师，副牧师	U5-2
curriculum *n.* 教学大纲，全部课程	U6-2
cutting-edge *adj.* 前沿的，尖端的	U7-1
darn *v.* 织补，缝补；*n.* 织补	U3-2
debut *vi.* 初次登台；*n.* 初次登台，开张	U5-1
debut *n.* 首次露面，初次登场；*vt.&vi.* 首次演出	U1-3
degradation *n.* 降解；降低	U2-2
deluxe *adj.* 豪华的，奢侈的，高级的	U7-2
denier *n.* 极少量；旦尼尔（测量纤维、尼龙或丝的光洁度的单位）	U2-2
denim *n.* 粗斜纹布，（复数）工作服，牛仔裤	U8-1
dermatitis *n.* ［U］皮(肤)炎	U1-1
destine *vt.* 命定，注定；指定	U6-2
detergent *n.* 洗涤剂，净洗剂	U2-1
detrimental *adj.* 有害的，不利的 (to)	U1-1
differentiate *v.* 区分	U8-3
diffusion *n.* 扩散，传播	U8-3
digital *adj.* 数字的，数码的	U8-2
dimension *n.* 方面，维度	U8-3
diode *n.* 二极管	U8-2
discard *n.* 抛弃，被抛弃的东西或人；*vt.* 抛弃，放弃，丢弃；*vi.* 放弃	U5-1
disperse *n.* 分散剂；*vt.* 驱散，解散，疏散；传播，散发	U1-1
dissect *v.* 解剖，详细分析	U8-3
distaff *n.* 卷线杆；女人；女红；女子关心的事	U3-3
distributor *n.* 销售者；批发商	U7-1
diverse *adj.* 多种多样的	U8-3
dominance *n.* 支配，控制；统治，优势	U1-2
downturn *n.* 低迷时期	U8-1

drape *vt.* 立体裁剪；*n.* （面料的）悬垂	U6-2
drapery *n.* （总称）布匹；[U] 纺织品	U1-1
drastically *adv.* 大大地，彻底；激烈地	U1-1
draw *vt.* 牵引，牵伸，拉伸	U2-3
drawing *n.* 牵伸	U3-3
drawing-in *n.* 穿经	U4-1
durability *n.* 耐久性，持久性	U2-1
dye *vt.* 染色；*n.* 染料	U2-3
dyeing *n.* 染色；*adj.* 染色的	U1-1；U2-1
eclipse *v.* 使……黯然失色	U8-1
elaborate *adj.* 精巧的，详尽的，复杂的	U1-1
elasticity *n.* 弹性，弹性学，弹力，伸缩力	U2-1
eliminate *v.* 消除，清除；排除（……的可能性）；淘汰，出局（常用于被动句）；杀害，消灭	U4-1；U5-1
elongation *n.* 伸长	U2-2
embankment *n.* [U] 筑堤；（河、海的）堤岸，（铁路的）路堤	U1-1
embossed *adj.* 浮雕式压花的	U4-3
embrace *vi.* 拥抱；*vt.* 拥抱，接受；包含，包围	U1-3
embroider *vt.* 在……上刺绣，给……修饰；*vi.* 刺绣，修饰，镶边	U1-3
embroidery *n.* 刺绣，刺绣品	U1-2
encompass *v.* 围绕，包括，完成	U8-3
endorse *v.* 赞同，支持	U8-3
end-user *n.* 最终用户，实际用户	U1-1；U7-1
engorged *adj.* 塞得满满的，过饱的	U5-3
entourage *n.* 随行，随从人员；周围，环境	U1-2
enviable *adj.* 令人羡慕的	U6-3
epidemic *n.* 流行，蔓延；流行病，传染病；*adj.* 流行的，传染性的	U1-2
episode *n.* （人生的）一段经历；（小说的）片段；（电视剧的）一集	U6-3
epoch *n.* （新）时代，（新）时期；重要时期，值得纪念的事件（或日期）	U1-2
eponymous *adj.* 齐名的；使得名的	U5-3
equalizer *n.* 均衡器，补偿器，平衡杆	U1-3
establishment *n.* 建立的机构；公司；会团；学校；机关；企业	U6-1
esthetic *adj.* 美的，美学的；美感的	U6-1
ethos *n.* 民族精神，时代思潮，社会风气，气质	U1-3
Eurasia *n.* 欧亚大陆	U1-2
exclusive *adj.* 排外的，唯一的，奢华的	U8-3

exclusivity *n.* 排外，排外主义；排他性，排他主义	U7-3
executive *n.* 行政主管	U8-1
exemplify *vt.* 例证，示范	U8-2
Exodus *n.* 《出埃及记》（《圣经》第二卷）	U3-2
exotic *n.* 外来物，外来的人；*adj.* 具有异国情调的	U6-3
extensibility *n.* 伸长性，延伸性，展开性	U2-1
extension *n.* 伸长	U2-2
exterior *n.* 外部，外面，表面，外形，外观	U7-3
extract *vt.* 提取，选取，摘录；*n.* 精华，提取物，摘录	U1-2
exuberant *adj.* 精力充沛的，热情洋溢的	U6-2
eyewear *n.* 眼镜	U6-1
fabric *n.* 织物，纤维，布料	U1-1；U2-1；U6-1；U7-2
fallout *n.* 后果，余波	U6-3
far-fetched *adj.* 强词夺理的，牵强附会的	U7-3
fawn-coloured *adj.* 淡黄褐色的	U5-3
Fayum *n.* 法雍（埃及一地区）	U3-2
feature *n.* 特征，特点；容貌，面貌；（期刊的）特辑；故事片； *vt.* 使有特色；描写……的特征；以……为号召物； *vi.* 起主要作用；作重要角色	U3-1；U4-1
felt *n.* 毛毡，毡制品；*vt.* 把……制成毡；用毡覆盖；*vi.* 毡合、毡化	U1-1；U2-2
felting *n.* 毡化	U1-1
fiber *n.* 纤维	U2-1
fierce *adj.* （竞争等）激烈的，强烈的；(人或动物)凶猛的，凶残的	U6-2
filament *n.* 丝，长丝，细丝，长纤维	U2-1；U2-3；U3-1
filling *n.* （织品的）纬纱	U4-1
filter *n.* 滤色镜，过滤器	U8-2
filtering *n.* 过滤，过滤作用	U1-1
fingering *n.* 绒线	U3-1
finish *v.* 后整理	U1-1
finishing *n.* 后整理，织物整理	U2-1
flax *n.* 亚麻	U2-2
flock *n.* 植绒	U4-3
fluctuate *vi.* 波动	U2-3
forager *n.* 抢劫者；强征队员	U3-2
forge *v.* （尤指努力地）生产，制造	U7-1
form *n.* 构造，形态	U6-2

fraction *n.* 小部分，碎片，片断		U4-1
fragment *n.* 碎片，碎屑；片段；*vt.&vi.* （使）成碎片，（使）分裂		U1-2
frame *n.* 框架，结构；*adj.* 有木架的，有框架的；*vt.* 设计，建造，陷害；*vi.* 有成功希望		U5-2
frayed *adj.* 磨损的		U5-3
freelance *vi.* 当自由作家（或演员等）；*n.* 自由作家（或演员等）；*adj.* 自由作家（或演员等）的，独立的；*adv.* 作为自由作家（或演员等）；独立地		U6-1
fringe *n.* 流苏，毛边，边缘；次要，额外补贴；*vt.* 用流苏修饰，镶边		U4-2
frosty *adj.* 严寒的，霜冻的		U8-1
functionality *n.* 功能，功能性		U8-2
furtive *adj.* （人）偷偷摸摸的，鬼鬼祟祟的；（行动）秘密的		U7-2
futuristic *adj.* 未来的，未来主义的		U7-2
gallon *n.* 加仑（液量单位，美制等于 3.785 升，英制等于 4.546 升）		U1-1
garment *n.* 衣服，服装，服饰；*vt.* 给……穿衣服	U1-3；	U6-1
garner *v.* 〈诗〉贮藏，积累		U7-1
gasket *n.* 垫片		U8-2
gauge *n.* 隔距；机号	U3-1；	U5-1
gear *n.* 衣服；齿轮；排挡；传动装置；*v.* 调整	U5-3；	U8-3
Genesis n.《创世纪》（基督教《圣经》的首卷）［略作 Ge*n.*］		U3-2
georgic *n.* 田园诗		U1-2
geotextile *n.* 土工织物		U1-1
giveaway *n.* （招徕顾客的）赠品		U7-1
glamorous *adj.* 富有魅力的，迷人的	U1-3；	U6-2
glucose *n.* 葡萄糖		U2-2
gossip *n.* 绯闻		U6-3
gradient *adj.* 倾斜的		U5-3
grant *n.* 助学金，补助金		U6-2
gray *adj.* 灰色的，暗淡的		U8-1
grill *n.* （炉具的）烤架，焙盘		U7-3
gripper *n.* 夹子，片梭；夹纱器		U4-2
guide *n.* 导丝器		U2-2
haberdashery *n.* 男子服饰用品店		U4-3
hail *vt.* 赞扬，称颂		U6-3
halt *n. vt.&vi.* 暂停		U6-3

hand *n.* 手感（同 handling; handle; hand feel; hand touch）	U2-2
handcraft *n.* 手工艺，手工	U3-1
handful *n.* 少数人（或事物），一把（的量）；*adj.* 一把；少数	U6-2
hands-on *n.* 实际动手经验	U6-2
hardwearing *adj.* 耐磨的	U2-3
harness *n.* ［印，纺］综绕；（提花机上的）通丝	U4-1
harshness *n.* 严厉，苛求	U6-3
heddle *n.* 综线	U4-1
high-fashion *adj.* （尤指女式服装）最新款式的，最新式样的	U7-3
high-performance *adj.* 高性能的	U8-2
high-profile *adj.* 引人注目的	U6-3
hinge *n.* 铰链；*v.* 依情况而定	U8-3
hip *adj.* (hipper, hippest)（非正式）时髦的，时尚的	U7-1
hitch *v.* 被挂住，被钩住	U7-1
homage *n.* 致敬	U5-3
hominid *n.* 原始人类	U8-1
Homo sapiens *n.* 智人（现代人的学名）；人类	U3-2
hood *n.* 头巾，兜帽	U8-2
horizontal *adj.* 水平的	U5-3
hose *n.* 软管，长筒袜，男性穿的紧身裤；*vt.* 用软管浇水；痛打	U5-2；U8-2
hosiery *n.* 针织品，袜类	U5-2；U6-1
hub *n.* 中心	U8-3
humanitarian *adj.* 人道主义者的，慈善家的	U7-3
humidity *n.* 湿度	U3-1
humiliation *n.* 羞辱，耻辱	U6-3
hurricane *n.* 飓风	U8-2
hydrophobic *adj.* 憎水性，疏水性	U2-3
hysteria *n.* 癔症；（特指女人的）歇斯底里；病态的兴奋	U7-3
icon *n.* 偶像，崇拜对象	U6-3
identification *n.* 检验，辨认	U2-3
illegitimate *adv.* 私生的，非法的	U6-3
illuminate *vt.* 照明；阐释，说明	U8-2
immaculate *adj.* 毫无瑕疵的，无缺点的	U7-1
impetus *n.* 动力，促进，冲力	U3-2
implant *vt.* 埋置；灌输，注入；种植，［医］移植	U1-1
incorporate *vt.* 合并	U8-2

incorporate *v.*	结合，合并，收编	U7-1
indicate *vt.*	表明，指示，预示，象征	U5-2
Indus *n.*	印度河（南亚河流）	U3-2
infamous *adj.*	臭名昭著的，声名狼藉的	U6-2
innovation *n.*	创新，革新，改革	U1-2
innovative *adj.*	革新的，有改革精神的，引进新观念的	U6-3
integral *adj.*	构成整体所必需的，不可缺的（+to）	U1-1；U7-1
integral *adj.*	积分的，部分的，整体的；*n.* （数学）积分，部分，完整	U5-1
interlace *vt.*	使交织，使组合；*vi.* 交错，组合，穿插	U1-1
interlock *v.*	（使）连锁，（使）联结，（使）连扣	U1-1
intern *n.*	实习生	U6-2
internship *n.*	实习生的职位；实习期	U6-1
in-tray *n.*	（办公室使用的）公文格，收文篮	U5-3
intricate *adj.*	复杂的，错综的	U5-2
inventory *n.*	盘存，存货；（财产等的）清单，报表；（商品的）目录	U7-3
ironic *adj.*	讽刺的，说反话的	U8-1
ironing *n.*	熨烫	U2-1
jeopardize *vt.*	危害，使陷危地	U5-1
journalism *n.*	新闻学	U6-2
juicer *n.*	榨果汁器	U8-1
jumper *n.*	妇女穿的套头外衣，连兜头帽的皮外衣，套头衫	U2-1
jute *n.*	黄麻	U2-2
keratin *n.*	角蛋白	U2-2
Kevlar *n.*	纤维 B（一种质地牢固重量轻的合成纤维）；［商标］凯夫拉尔	U8-2
kimono *n.*	和服	U7-2
knack *n.*	技能，本领	U6-3
knitted *adj.*	针织的，编织的	U2-1
knitting *n.*	［U］编织；（总称）编织物	U1-1
knotless *adj.*	无结的	U3-2
lace *n.*	［U］花边，蕾丝，饰带；［C］鞋带；带子；*v.* 穿带子于，用带系（+up）；用花边等装饰	U1-1
lacing *n.*	花边织法；结带；镶边；饰带，花边	U1-1
lap *n.*	一圈，膝盖；下摆；山坳；*vt.* 使重叠；拍打；包围；*vi.* 重叠；轻拍；围住	U3-3
lapel *n.*	（西服上衣或夹克的）翻领	U5-3
launch *vt.*	发起	U8-2
layoff *n.*	临时解雇，停止操作	U8-1

lean *v.*　倾斜，倚靠	U8-1
legacy *n.*　遗留之物，遗产	U6-3
legion *n.*　众多，大批，无数	U7-1
legitimacy *n.*　合法性；正统性；嫡系	U7-3
lengthwise *adj.*　纵长的；*adv.*　纵长地	U3-3
lift *v.*　正式取消，结束	U8-1
linen *n.*　[U] 亚麻布，亚麻线（纱）；亚麻布制品（如床单、桌巾、内衣等）	U1-1；U2-2；U2-3；U8-2
longitudinally *adv.*　纵向地	U2-3
loom *n.*　[C] 织布机；[U] 织造术；*vt.*　在织布机上织	U1-1；U1-2
loop *n.*（线，铁丝等绕成的）圈，环	U1-1
loops *n.*　线圈；*vt.*　使……成环	U5-1
loose *adj.*　宽松的	U6-3
lubricate *vt.*　使滑润，给……上润滑油	U1-1
luster *n.*　光彩，光泽	U2-2；U4-3
Lycra　莱卡，人造弹性纤维品牌	U4-3
mainstream *n.*（思想或行为的）主流；主要倾向，主要趋势	U5-3
mannequin *n.*　人体模型，时装模特	U6-2
mansion *n.*　豪宅，公馆	U6-3
mantra *n.*　颂歌，圣歌；咒语	U7-2
manual *adj.*　手的；手制的，手工的；*n.*　手册；指南	U4-1
manufacturer *n.*　制造商	U4-1
mat *n.*　丛，簇，团（+of）；地席，草席；垫子	U1-1
maternity *n.*　孕妇装	U6-1
matting *n.*　席子；编席的原料；*v.*　纠缠在一起；铺席于……上；使……无光泽	U3-2
mentality *n.*　脑力，智力；精神；心理，意识；思想	U7-3
mentally *n.*　精神力，心理状态	U8-1
microprocessor *n.*　微处理器	U4-1
microscope *n.*　显微镜	U2-3
millennia *n.*　(*pl.*) millennium 的复数	U1-2
millennium *n.*　一千年	U1-2；U1-3
millimeter *n.*　毫米	U4-1
mindset *n.*　心态；倾向，习惯	U1-1
minimalist *n.*　极简抽象派艺术家，极简派音乐家，保守派；*adj.*　极简抽象艺术的，极简抽象风格的；最低限度的	U5-3
mirror *vt.*　反映	U6-3

miscellaneous *adj.*	混杂的，五花八门的，各种各样的；多才多艺的	U1-1
mistress *n.*	情妇	U6-3
module *n.*	组件，单元，模块	U8-2
mohair *n.*	马海毛，安哥拉羊毛	U2-2；U3-1
monochrome *adj.*	单色的	U6-3
monopoly *n.*	垄断，独占，控制	U1-2
morphine *n.*	吗啡	U6-3
moss *n.*	苔藓，藓沼，泥炭沼	U5-3
moth *n.*	飞蛾，蛾子	U1-2
mould *n.*	铸模，模型	U5-3
multiple *adj.*	许多的；*n.* 倍数	U8-3
muse *n.*	缪斯；/希神/（文艺、美术、音乐等的）女神；灵感	U6-3
muted *adj.*	（颜色）柔和的，不鲜艳的	U5-3
myriad *n.*	无数，大量 (+of)；*adj.* 无数的，大量的；各种各样都有的	U1-1
naked *adj.*	裸体的	U6-3
nanotechnology *n.*	纳米技术	U8-2
nap *n.*	绒	U4-3
Navajo *n.*	纳瓦霍人（美国最大的印第安部落）	U3-2
needle *n.*	织针	U5-2
Neolithic *adj.*	［古］新石器时代的；早先的	U3-2
Nordic *adj.*	北欧人的	U7-2
nozzle *n.*	喷嘴	U4-2
nun *n.*	修女	U6-3
nylon *n.*	［U］尼龙，(*pl.*) 尼龙长袜	U1-1
obscure *vt.*	使……模糊不清，使隐晦，使费解；掩盖； *adj.* 不易看清的，暗淡的；费解的，难以理解的	U6-3
ocher *n.*	赭色	U8-1
ochre *n.*	赭石；赭色，黄褐色	U5-3
operative *adj.*	有效的；运转着的；从事生产劳动的；*n.* 侦探；技工	U3-2
optical *adj.*	光学的	U8-2
option *n.*	选择	U8-1
oracle *n.*	神使，先知，预言者；大智者；（谑）圣人，哲人	U7-3
orchestrate *v.*	为（管弦乐队）谱写音乐；使和谐地结合起来	U7-1
ornamentation *n.*	装饰品，装饰	U6-1
ornate *adj.*	装饰华丽的	U6-3
outerwear *n.*	（总称）外衣，外套	U6-1

outfit *n.* 装置；（一定场合下穿的）全套衣装	U7-3
outlet *n.* 零售网点，经销点，专卖店；（美）（通常坐落在市郊的）购物中心，廉价商品销售中心	U7-1
outrageous *adj.* 令人吃惊的，出人意料的；无礼的，令人无法容忍的	U6-2
oversee *vt.* 监视，监督，管理，看管	U6-1
pale *adj.* 苍白的，灰白的；（颜色）淡的	U1-1
Palestine *n.* 巴勒斯坦（西南亚地区名）	U3-2
palette *n.* 调色板，颜料	U5-3
panel *n.* 衣片	U5-1
panne *n.* 平绒	U4-3
para-aramid *n.* 对位芳纶	U8-2
parachute *n.* 降落伞	U1-1
paramount *adj.* 最高的，至上的，首要的	U7-1
parish *n.* 教区	U5-2
partially *adv.* 部分地	U8-1
pashmina *n.* 开司米亚羊毛	U1-1
passionate *adj.* 充满激情的	U6-2
patch *vt.* 修补，补缀，拼凑；*n.* 补丁，碎片	U1-3
patternmaker *n.* 打样师；制模师	U6-1
pectic *adj.* 果胶的，黏胶质的	U2-2
people *vt.* 居住于；把……挤满人，住满居民	U6-3
peripheral *adj.* 外围的，不重要的	U8-3
perpetuate *vt.* 使永存，保存	U1-3
perseverance *n.* 坚持不懈，不屈不挠	U5-2
Persian *adj.* 波斯的，波斯人的，波斯语的；*n.* 波斯人，波斯语	U1-2
Phaedra *n.* ［希神］菲德拉	U1-2
phase *n.* 阶段，时期；*v.* 逐步执行	U8-2
Phrygia *n.* 佛里吉亚（小亚细亚中部一古国）	U3-2
pick *n.* 纬纱；投梭	U4-1
pile *n.* 绒毛	U4-3
pill *vi.* 起球	U2-3
playsuit *n.* （妇女，儿童）运动衫，运动裤	U7-2
pleat *n.* 褶	U2-3
Pleistocene period *n.* ［地质］更新世时期	U3-2
plummet *vi.* 垂直落下，暴跌	U8-1
polarity *n.* 两极	U8-3

polarization *n.*	两极化	U8-3
polarized *adj.*	两极化的	U8-3
polyacrylic *n.*	聚丙烯酸化合物，腈纶	U2-1
polyamide *n.*	聚酰胺，尼龙，锦纶，耐纶	U2-1；U8-2
polyester *n.*	［化］聚酯；涤纶	U1-1；U2-1
polyethylene *n.*	［高分子］聚乙烯	U8-2
polymer *n.*	聚合物	U2-1；U2-3；U8-2
portfolio *n.*	（艺术家等的）代表作选辑；文件夹，卷宗夹	U6-1
potent *adj.*	强有力的，有效的	U1-3
potential *adj.*	潜在的，可能的；*n.* 潜能，可能性，电势	U5-1
precarious *adj.*	危险的，不稳定的	U5-2
precede *vt.*	领先，优于，在……之前；*vi.* 领先，在前面	U5-2
predominantly *adv.*	占主导地位地，占优势地，显著地	U1-1
preforming *n.*	预成型，压片	U8-2
preliminary *adj.*	初步的，预备的；*n.* 初步，开端，预备；预考；预赛	U6-1
premise *n.*	（理由等的）前提，根据，缘起部分	U7-1
preppy *adj.*	预备学校学生的，（尤指在衣着、举止等方面）像预备学校学生的	U5-3
present *vi.*	举枪瞄准；*adj.* 现在的，出席的；*n.* 现在，礼物，瞄准	U5-2
prestigious *adj.*	受尊敬的，有名望的，有威信的	U6-2
prêt-a-porter *n.*	现成服装	U8-3
pretension *n.*	虚饰，假装，做作，虚荣，自负	U7-3
prevalence *n.*	流行，盛行，普遍	U1-2
prevalent *adj.*	流行的，盛行的，普遍的 (+among/in)	U1-1
preview *n.*	预观，预映，试映，预演，试演；（展览会的）预展；预习	U7-1
prewinder *n.*	预络纱机	U4-1
primitive *adj.*	原始的，早期的；简单的，粗糙的；质朴的，自然的；*n.* 原始人，原始事物	U1-2
printing *n.*	印花，印花工艺，印刷	U2-1
priority *n.*	重点，优先（权）	U1-3
procure *vt.*	（努力）取得，获得；采办；为……获得	U6-1
procurement *n.*	采购	U7-3
profile *n.*	简介，概况；外形，轮廓；*vt.* 描绘……的轮廓，为……写传略	U1-3
profound *adj.*	深远的，深厚的，深重的	U8-1
projectile *n.*	片梭	U4-2
proportion *n.*	大小比例	U6-2；U8-3
protein *n.*	蛋白质；*adj.* 蛋白质的	U2-1

prototype *n.* 原型，样板；标准，模范	U6-1
psychological *adj.* 心理的	U8-1
puckering *n.* 皱纹，褶皱	U4-2
pulp *n.* 果肉；纸浆	U8-2
quota *n.* 配额，定额，限额	U8-1
rack *n.* 行李架，衣架	U8-1
rage *n.* 风行，狂热；狂怒，盛怒；*vi.* 盛行，流行；狂怒，大怒	U1-3
ramie *n.* 苎麻，苎麻纤维	U2-2
razor *n.* 剃刀，剃须刀，刮脸刀	U6-3
recapture *vt.* 重新获得，夺回，收复；*n.* 重占，夺回，收复	U1-3
reclaim *v.* 缫丝，回收	U2-2
recount *vt.* 详述，列举；*n.* 重新计算	U1-2
redundant *adj.* 过多的，冗余的，冗长的（文章等）；累赘的；丰富的（食物等）	U7-2
reel *v.* 缫丝	U2-2
refinement *n.* 精炼，提纯，净化；改良品；极致；优雅、高贵的动作	U4-1
reign *vi.* 盛行，支配，统治；*n.* 盛行，支配；君王统治，在位期	U1-3
release *n.* 发表；发售（物）；（影片的）发行上映	U7-1
religiously *adv.* 笃信地，虔诚地	U6-2
resiliency *n.* 弹性	U2-2
resistance *n.* 阻抗性，抵抗，抵抗力，阻力，电阻	U2-1
restructure *v.* 重组	U8-1
resume *n.* 简历，履历；*vt.&vi.* 重新开始，重新获得	U6-2
retail *n.&vt.* 零售	U8-3
retention *n.* 保留，保持，维持	U7-1
retrench *vt.&vi.* 减少，削减（经费，开支等）	U1-3
retro *adj.* 复古的，怀旧的；*n.*（服装式样等）重新流行，复旧，怀旧	U5-3
revenue *n.* 收入	U6-3
revitalize *vt.* 使振兴，使复兴，使恢复	U1-3
revive *vt.* 唤醒，使……重生	U8-2
revolution *n.* 革命，剧烈的变革；回转，绕转，旋转，转数；周期；一转	U4-1
revolutionise (or revolutionize) *vt.* 使彻底变革	U6-3
rickshaw *n.* 人力车，黄包车	U7-2
rotting *adj.* 腐烂的	U6-2
roving *n.* 粗纱	U3-1
ruffle *n.* 褶边，褶边状物	U7-2
rugged *adj.* 结实的，坚固的	U1-3

rumour *n.*	谣言	U6-3
scarf *n.*	围巾，头巾	U6-1
scheme *n.*	计划，方案	U8-2
Scotch blackface *n.*	苏格兰黑面羊	U3-2
scout *vt.*	物色（优秀音乐家、艺术家等），寻找	U6-2
scrap *n.*	碎片，小块；*adj.* 零碎的	U1-2
scrap *n.*	碎片，残余物；*adj.* 废弃的，零碎的；*vt.* 废弃，使解体，拆毁；*vi.* 吵架	U5-1
seam *n.*	缝，接缝；*vt.* 缝合，结合；*vi.* 裂开，产生裂缝	U5-2；U6-3
seamstress *n.*	女裁缝	U6-2
segment *n.*	部分	U8-2
selective *adj.*	精心选择的	U6-2
selvage *n.*	织边，镶边，布边	U4-2
send-off *n.*	送别，送行	U7-3
sensitivity *n.*	灵敏性	U2-3
sensor *n.*	传感器，探测器	U8-2
sericin *n.*	丝胶	U2-2
sericulture *n.*	养蚕（业）	U1-2
serviceable *adj.*	有用的；可供使用的	U3-1
sew *vt.*	缝，缝合，缝纫	U6-1
shade *n.*	颜色，色调，色泽，色光	U2-1
shaft *n.*	柄，轴；矛，箭；*vt.* 给……装上杆柄	U4-1
share *n.*	份额，股份；*v.* 分享，分担	U2-3
shear *v.*	剪	U4-3
sheer *adj.*	全然的	U8-1
shoot *n.*	摄影，拍电影	U7-3
simultaneously *adv.*	同时地	U5-1
sizing *n.*	上浆，上胶，胶料，填料	U4-2
skein *n.*	绞纱，一束（线或纱）	U2-2；U3-1
sketch *vt.*	画草图；草拟；*n.* 速写，素描；略图，草图，粗样，草稿	U6-1；U7-3
sleek *adj.*	非常时髦的，豪华的；兴旺的	U7-1
slick *adj.*	光滑的，滑溜的；熟练的，灵巧的	U4-1
slippery *adj.*	光滑的	U2-3
slitting *n.*	切口，切缝，纵切，纵裂	U4-2
sliver *n.*	棉条	U3-1
solicitor *n.*	咨询律师，［英］出席初级法院的诉状律师，事务律师；［美］法务官	U5-3
solidarity *n.*	团结，齐心协力	U6-3

solubility *n.*　可溶性	U2-3
sophistication *n.*　复杂，老于世故，有教养	U5-1
spandex *n.*　（作腰带、泳衣用的）弹性人造纤维（织物）、氨纶	U1-1
spatial *adj.*　空间的	U8-3
spectrum *n.*　范围，幅度，系列，光谱	U1-3
speculation *n.*　推测	U6-3
spin *v.*　纺纱，纺织	U2-2
spindle *n.*　纺锤，锭子；细长的人或物；*adj.*　锭子的，锭子似的；细长的	U3-2
spinneret *n.*　纺丝头，喷丝头	U2-1；U2-3
spinning *n.*　纺纱	U2-3；U3-1；U3-2
spool *vt.*　缠绕；卷在线轴上；*n.*　线轴；缠线框	U3-3
spool valve *n.*　短管阀	U3-3
spouse *n.*　配偶，夫，妻；夫妇	U7-2
spurn *vt.&vi.*　摒弃，拒绝，藐视	U1-3
staidness *n.*　认真，沉着	U7-1
stain *n.* 污点	U2-3
stained *adj.*　玷污的，褪色的	U6-3
staple *n.*　主要产品，重要特色；订书钉；纤维，短纤维，毛束，纤维长度，U 形钉；*adj.*　最基本的，最重要的；*vt.*　用订书机装订	U1-3；U5-3；U2-1；U3-1
starch *n.*　［U］淀粉；淀粉类食物；（浆衣服等用的）淀粉浆	U1-1
state-of-the-art *adj.*　最新型的，最先进的，顶尖水准的，使用了最先进技术的	U4-1
stitch *n.*　一针，针脚，线迹；［C；U］针法；编结法；*v.*　缝，绣，编结 (+up)	U1-1；U3-1
stitches *n.*　线圈	U5-1
stocking *n.*　长筒丝袜，丝袜	U5-2
straggly *adj.*　脱离行列的，落后的，蔓延的，散乱的	U5-3
streak *n.*　纹理，条纹，斑纹，条痕，条层，色条，色线	U4-2
strength *n.*　强力，强度	U2-1
striate *adj.*　有条纹的	U2-3
stripe *n.*　条纹	U5-3
strut *v.*　大摇大摆地走，趾高气扬地走	U7-1
stud *vt.*　镶嵌，点缀；*n.*　大头钉，饰钉，金属扣	U1-3
stuffy *adj.*　闷热的，不通气的；古板的，保守的；枯燥无味的；一本正经的	U5-3
stylist *n.*　造型师	U6-2
substandard *adj.*　标准以下的	U7-1
suit *n.*　（一套）衣服	U6-1

summon *vt.*	传讯（出庭），传唤，召集	U6-3
sumptuous *adj.*	奢侈的，豪华的；高价的	U7-2
supplier *n.*	供应者，供货商	U6-1
surplus *adj.*	剩余的，多余的；*n.* 剩余，多余	U1-3
swatch *n.*	（布料等的）样品	U8-1
sweater *n.*	毛衣	U5-1
sweatshop *n.*	（非正式）血汗工厂	U7-3
swoon *vi.*	晕厥，昏倒；心醉神迷，神魂颠倒；*n.* 晕厥，昏倒，狂喜	U1-3
synthetic *adj.*	合成的，人造的，综合的	U1-1
Syria *n.*	叙利亚共和国	U3-2
tabby *n.*	平纹；斑猫；长舌妇；*adj.* 起波纹的；有斑纹的；*vt.* 使起波纹	U3-2
tag *n.*	标签；附笺，贴纸	U7-3
tailoring *n.*	裁缝业，成衣业	U1-1
tangle *vt.&vi.*	纠结，乱成一团	U1-1
tangle *vt.*	缠结	U2-3
temporal *adj.*	时间的	U8-3
tenacity *n.*	韧性，强度	U2-2
tension *n.*	强度	U3-1
tex *n.*	支数（纱线疏密程度的一种单位，每千米纱线或者纤维所具有的质量克数）	U3-1
textile *n.*	纺织品，织物	U1-1；U2-1
texturize *vt.*	变形处理	U2-3
thermochromic *adj.*	热色性的	U8-2
thermoplastic *adj.*	热塑性的	U2-3
tornado *n.*	龙卷风	U8-2
touch *n.*	装点，润色	U6-1
transfer *vt.*	转移，调任	U8-2
traumatize *vt.*	使受损伤	U8-1
trendy *adj.*	时髦的，赶时髦的，追随时髦的	U6-2；U7-1
trim *n.*	布边	U5-1
trot *v.*	（人）匆忙地走，快步走	U7-2
Tsar *n.*	沙皇	U6-3
tufting *n.*	丝线法	U4-3
twill *n.*	斜纹织物；*vt.* 把……织成斜纹；*adj.* 斜纹织物的	U1-1
twist *vt.*	捻，捻合	U3-1
unabated *adj.*	不衰退的，不减弱的	U8-3

underarm *adj.* 腋下的，手臂内侧的；*adv.* （接球时）用低手	U5-1
underground *adj.* 反传统的，反现存体制的，（艺术等）先锋派的	U7-3
understated *adj.* 朴素的，简朴的；轻描淡写的；有节制的；低调的	U5-3
under-the-radar *adj.* 不引人注目的，低调的	U5-3
unthinkable *adj.* 难以想象的，不可思议的，难以置信的；不可能的	U1-1
upholster *vt.* 为（沙发、椅子等）装上垫子（或套子、弹簧等）(+in/with)；用（挂毯、家具等）布置（房间）；装潢	U1-1
upholstery *n.* 家具装饰用品	U4-3
up-to-date *adj.* 最新（式）的；现代化的，尖端的；直到最近的	U4-1
usability *n.* 可用；合用；可用性	U1-1
utterly *adv.* 完全地，彻底地，绝对地，十足地	U6-2
vague *adj.* 模糊的；茫然的	U8-3
vapour *n.* 水蒸气，雾气	U8-2
varying *adj.* 变化的，不同的	U6-1
velour *n.* 天鹅绒	U4-3
velveteen *n.* 平绒	U4-3
versatile *adj.* 通用的，多功能的，多才多艺的	U1-3
versatility *n.* ［U］多用途，多功能；多才多艺	U1-1；U5-3
vest *n.* ［美］背心，马甲，防护背心	U1-1
viable *adj.* 能生存的，可行的	U8-1
vibe *n.* 感应，感觉	U5-3
virile *adj.* 男性的，有男子气概的，强壮的	U1-3
viscose *n.* 黏胶液，黏胶纤维	U2-1；U4-3
visionary *adj.* 有远见的	U6-3
vulnerable *adj.* 脆弱的，敏感的；易受伤的，易受责难的	U6-3
waistband *n.* 腰带，裤带	U7-2
wardrobe *n.* 衣橱，衣柜；全部服装	U1-3；U5-3；U6-3
warp *n.* ［纺］［the+S］（棉布的）经线	U1-1；U4-1；U4-3
water-resistant *adj.* 抗水的，防水的	U4-2
wax *n.* ［U］蜡，蜂蜡，石蜡，蜡状物；*adj.* 蜡制的；*vt.* 给……上蜡	U1-1
wearable *n.* 衣服，服装	U7-1
weary *adj.* 疲倦的，厌烦的；*vt.* 使疲倦，使厌烦；*vi.* 疲倦，厌烦	U1-3
weave *vt.* 织，编，编制	U1-1
weaver *n.* 织工	U3-1
weaving 编，织	U1-1；U3-2
weft *n.* ［纺］［the+S］纬线，纬纱；织品；薄云层	U1-1；U4-1；U4-3

welder *n.* 焊工	U1-1
wholesaler *n.* 批发商	U6-1
whorl *n.* 螺纹；轮生体，涡；*vt.* 使成涡漩；*vi.* 盘旋	U3-2
wind *vt.* 缠绕；卷在线轴上	U3-3
winding *n.* 络纱	U4-1
wind-up *n.* 卷绕	U2-3
woodblock *n.* 木板，木块；［印］木版，木刻（画）	U1-1
wrinkle *n.* 皱纹，褶皱；*v.* 起皱	U2-1
yarn *n.* 纱，线	U1-1；U2-1；U4-1
yearn *v.* 渴望，盼望	U8-1
Zagros Mountains *n.* 扎格罗斯山脉（伊朗西南部山脉）	U3-2
zoom *n.* （摄影）可变焦距镜头	U7-2

Acknowledgements

We are very grateful to the following sources:

[1] http://wenku.baidu.com/view/d56226543c1ec5da50e2706f.html

[2] http://www.teonline.com/knowledge-centre/spinning.html

[3] http://www.ehow.com/about_5476382_staple-yarn.html

[4] http://wenku.baidu.com/view/0721c272f242336c1eb95e0f.html

[5] http://en.wikipedia.org/wiki/Nonwoven_fabric

[6] http://www.textileschool.com/School/Fabrics/NonWovenFabrics.aspx

[7] http://www.cottoninc.com/Cotton-Nonwoven-News/Fibers-for-Nonwovens/

[8] http://www.engr.utk.edu/mse/Textiles/Wet%20Laid%20Nonwovens.htm

[9] http://www.textileworldasia.com/Articles/2010/March/Recent_Developmentsx_Weaving_Technology.html#language

[10] http://www.teonline.com/knowledge-centre/industrial-knitting-process.html

[11] http://www.p2pays.org/ref/11/10023/weftwarpknitting.asp

[12] C. A. Lawrence, Fundamentals of spun yarn technology. CRC Press LLC, 2003.

[13] Yehia El Mogahzy, Understanding the Fiber-to-Yarn Conversion System Part Ⅱ: Yarn Characteristics.-

[14] http://www.fashion-era.com/velvets/velvet.htm#Choosing_Velvet_Pile_Types_of_Fabric

[15] http://www.ftchinese.com/story/001034729/ce

[16] Crewe, L. and Forester, Z (1993), "Markets, Design, and Local Agglomeration: The Role of the Small Independent Retailer in the Workings of the Fashion System", Environmetn and Planning D: Society and Space, 11:213-29

[17] Mintel International Group (2002a), Clothing Retailing in Europe—UK: Retail Intellignece, London: Mintel Itnerantional Group Ltd.

[18] Mintel International Group (2002b), Department and Variety Store Retailing:UK, European Retail Intelligence, Retail Intellignece, London: Mintel Itnerantional Group

[19] Breward, C., and D. Gibert, eds(2006), Fashion's World Cities, Oxford: Berg.

[20] http://en.wikipedia.org/wiki/History_of_silk

[21] http://blog.modelmanagement.com/2010/06/01/zara-and-hm-fast-fashion-on-demand/#comments

[22] http://www.chinadaily.com.cn/ezine/2007-08/01/content_5447100_2.htm

[23] http://fashion.telegraph.co.uk/news-features/TMG4449609/Online-shopping-fast-fashion-at-your- fingertips.html